Nimble with Numbers

Engaging Math Experiences to Enhance Number Sense and Promote Practice

Grades 5 and 6

Leigh Childs and Laura Choate

Dale Seymour Publications
Parsippany, New Jersey

Acknowledgments

Thanks to special friends and colleagues who helped field test these activities with their students and who gave valuable feedback and suggestions:

Polly Hill

Maryann Wickett

Thanks to super kids who provided a student perspective as they tried these activities:

Dan Hill

Jensa Mashek

Anya Newman

Jacob Newman

Luke Schwartz

Thanks to two wonderful, supportive husbands who contributed significantly to this three-year effort:

Mike Choate

Milt Schwartz

Executive Editor: Catherine Anderson
Project Editor: Patsy Norvell
Production/Manufacturing Director: Janet Yearian
Production/Manufacturing Manager: Karen Edmonds
Production/Manufacturing Coordinator: Joan Lee
Design Director: Phyllis Aycock
Design Manager: Jeff Kelly
Text design: Tani Hasegawa
Page composition: Joe Conte
Cover design and illustration: Ray Godfrey
Art: Stefani Sadler

Dale Seymour Publications
An imprint of Pearson Learning
299 Jefferson Road
Parsippany, NJ 07054
www.pearsonlearning.com
1-800-321-3106

ISBN 0-7690-2724-5

3 4 5 6 7 8 9 10-ML-05-04-03-02-01-00

Dale Seymour Publications®

Table of Contents

Decimals

Percents

Algebra Readiness

Blackline Masters

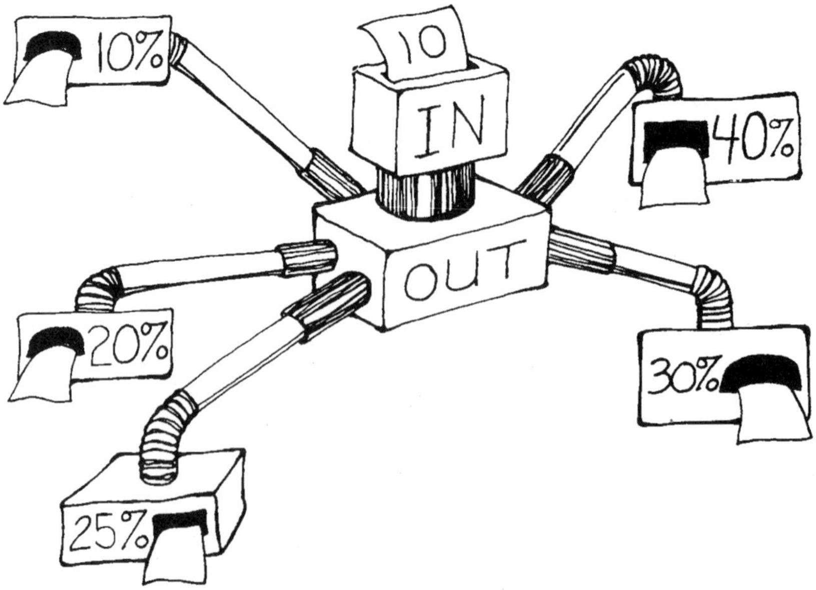

Introduction to *Nimble with Numbers*

Why This Book?

National recommendations for a meaning-centered, problem solving mathematics curriculum place new demands on teachers, students, and parents. Students need a facility with numbers and operations to achieve success in today's mathematics programs. Basics for students today require a broadening of the curriculum to include all areas of mathematics. Students are being asked to demonstrate proficiency not just in skills, but in problem solving, critical thinking, conceptual understanding, and performance tasks. Consequently, the time teachers devote to number topics must be thoughtful, selective, and efficient.

This book fulfills the need for high-quality, engaging math experiences that provide meaningful practice and further the development of operation and number sense. **These activities are designed to help students practice number and pre-algebra concepts previously taught for understanding in a variety of contexts.** Besides meeting the need for effective practice, *Nimble with Numbers:*

- Provides a variety of adaptable formats for essential practice
- Supplements and enhances homework assignments
- Encourages parent involvement in developing their children's proficiency with mental computation, fractions, decimals, percents, and pre-algebra topics
- Provides motivating and meaningful lessons for a substitute teacher

Criteria for Preferred Activities

For efficient use of time devoted to number-related topics, the book focuses on activities that are:

- Inviting (encourages participation)
- Engaging (maintains interest)
- Simple to learn
- Repeatable (able to reuse often and sustain interest)
- Open-ended, allowing multiple solutions
- Easy to prepare
- Easy to adapt for various levels
- Easy to vary for extended use

In addition, these activities:

- Require a problem-solving approach
- Improve basic skills
- Enhance number sense and operation sense
- Encourage strategic thinking
- Promote mathematical communication
- Promote positive attitudes toward mathematics as mathematical abilities improve

Planning Made Simple

Organization of Book

The activities of this book are divided into seven sections which cover high-priority number-related topics for fifth and sixth graders. The first three sections review basic facts and mental computation with all operations.

The fourth, fifth, and sixth sections reinforce fraction, decimal, and percent concepts with an effort to demonstrate the relationship between these three topics, particularly in the Percents section. When students seem confident with the first six sections, proceed to the more challenging seventh section. The Algebra Readiness section reinforces number theory topics and simple algebraic concepts and equations.

Each section begins with an overview and suggestions which highlight the activities and provide time-saving advice. The interactive activities identify the specific topic practiced (Topic), the objective (Object), the preferred grouping of participants (Groups), and the materials required (Materials). Activities conclude with "Making Connections" questions to promote reflection and help students make mathematical connections. "Tips" provide helpful implementation suggestions and variations. Needed blackline masters are included with the activity or in the Blackline Masters section at the end of the book.

The introductory section includes a Matrix of Activities. The repeatable Sponges and Games are listed alphabetically with corresponding information to facilitate their use.

Types of Activities

The book contains activities for whole group, small groups, pairs, and individuals. Each section provides:

- Sponges (S)
- Skill Checks (C)
- Games (G)
- Independent Activities (I)

Sponges

Sponges are enriching activities for soaking up spare moments. Use Sponges as warm-up or spare time activities with the whole class or with small groups. Sponges usually require little or no preparation and are short in duration (3–15 minutes). These appealing Sponges are repeatable and, once they become familiar, can be student-led.

Skill Checks

The Skill Checks in each section provide a way to show students' improvement to the parents, as well as to themselves. Each page is designed to be duplicated and cut in half, providing six comparative records for each student. Before answering the ten problems, students should respond to the starter task following the STOP sign. These starter tasks are intended to promote mental computation and build number sense. Some teachers believe their students perform better on the Skill Checks if the responses to the STOP task are shared and discussed before students solve the remaining ten problems. Most students will complete a Skill Check in 10 to 15 minutes. The concluding extension problem, labeled "GO ON," accommodates those students who finish early. We recommend that early finishers be encouraged to create similar problems for others to solve. By having students discuss their approaches and responses, teachers help students discover more efficient mental computation strategies.

Games

Initially a new Game might be modeled with the entire class, even though Games are intended to be played by pairs or small groups after the rules are understood. An excellent option is to share the Game with a few students who then teach the Game to others. To facilitate getting started, teachers may recommend some procedure for identifying the first player or pair. A recommended arrangement is to have players partner up and collaborate as "pair players" against another pair. Mathematical thinking and communication is enhanced as students collaborate to develop and share successful strategies. Some Games provide easier or more challenging versions. Most Games require approximately 20 to 45 minutes of playing time. Games are ideal for home use since they provide students with additional practice and reassure parents that the "basics" continue to be valued. When sending gameboards home, be sure to include the directions.

Games and Sponges

Games and Sponges provide students with a powerful vehicle for assessing their own mathematical abilities. During the Games, students receive immediate feedback that allows them to revise and to correct inefficient and inadequate practices. Sponges and Games differ from the Independent Activities since they usually need to be introduced by a leader.

Independent Activities

Independent Activity sheets provide computation practice and review number-related topics. These sheets are designed to encourage practice of many more number concepts than would seem apparent at first glance. Many Independent Activities allow multiple solutions. Most students will complete an Independent Activity sheet in 15 to 30 minutes. Independent Activities can be completed in class or sent home as homework. Some Independent Activities provide two versions to accommodate different levels of difficulty. In the Fractions and Decimals sections the two versions of some Independent Activities allow practice of addition and subtraction or multiplication and division. Many Independent Activities can be easily modified to provide additional practice.

Suggestions for Using *Nimble with Numbers*

Materials Tips

An effort has been made to minimize the materials needed. When appropriate, blackline masters are provided. The last section of the book contains more generic types of blackline masters. The six-sectioned spinner (p. 156) can substitute for a number cube or die. The blank spinner can be used for the specially marked cubes (.1 – .6 or +, +, +, –, –, –). A simple spinner, like the one shown, can be assembled using one of the blackline master spinner bases, a paper clip, and a pencil.

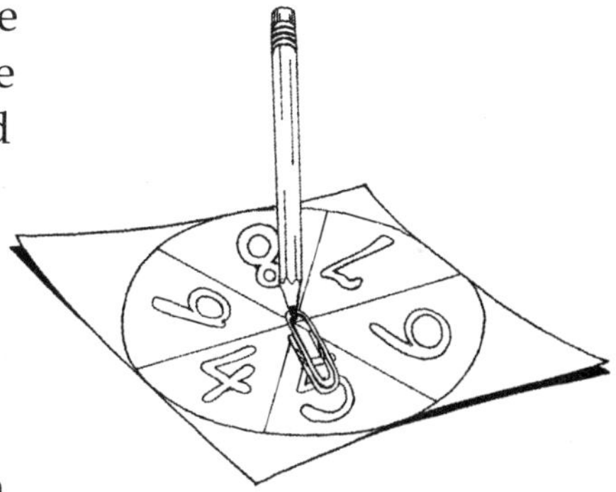

Many activities use the Digit Squares (p. 151). The familiar sets of 0–9 number tiles substitute well for Digit Squares. If not available, take time now to duplicate a Digit Square set on card stock for each student.

A few activities and class Sponges require Digit Cards (p. 150). Digit Card sets should also be duplicated on card stock. Teachers should cut two sets of Digit Cards apart, place them in an appropriate container (paper sack, coffee can, or margarine tub), and store in a handy place.

Various materials work as markers on gameboards—different types of beans, multi-colored cubes, buttons, counters, or transparent bingo chips (our preference due to the see-through feature). For some activities students will need scratch paper and pencils. It is assumed that an overhead projector is available, but a chalkboard or whiteboard may be substituted.

Recommended Uses

The repeatable nature of these activities makes them ideal for continued use at home. Encouraging students to use these activities at home serves a dual purpose: parents are able to assist their students in gaining competence with mental computation and important number concepts, and parents are reassured as they see the familiar basics practiced. To support your work in this area, we have included a parent letter and a list of helpful open-ended questions.

Besides being a source for more familiar homework, these activities offer a wide variety of classroom uses. The activities can be effectively used by substitute teachers, as rainy-day options, or for a change of pace. Many activities are short-term and require little or no preparation, making them ideal for soaking up spare moments at the end or beginning of a class period. They also work well as choices for center or menu activities. When students are absent from school, include these activities in independent work packets. You may package these activities in manila envelopes or self-closing transparent bags to facilitate frequent and easy checkout. To modify the activities and to accommodate the needs of your students, you may easily change the numbers, operations, and directions.

Getting the Most from These Activities

It is important to focus on increasing students' awareness of the mathematics being learned. To do this, pose open-ended questions that promote reflection, communication, and mathematical connections. To assist you with this vital task, "Making Connections" questions are included with each Sponge and Game.

Having students work together as pair players is of great value in increasing student confidence. While working this way, students have more opportunities to communicate strategies and to verbalize thinking. When asked to identify and to share their successful Game strategies verbally and in writing, students grow mathematically. Also it is worthwhile to ask students to improve these activities or to create different versions of high-interest games.

Good questions help students make sense of mathematics, build their confidence, and encourage mathematical thinking and communication. A list of helpful, sample questions appears on page 9. Since the teacher's or parent's response impacts learning, we have included suggestions for responding. Share this list with parents for their use as they assist students with these activities and other unfamiliar homework tasks. This list was created by Leigh Childs for parent workshops and for inclusion in the California Mathematics Council's *They're Counting on Us, A Parent's Guide to Mathematics Education.* We have adapted the list for use with this book.

Concluding Thought

We hope that by using these materials, your students will develop more positive feelings towards mathematics as they improve their mathematical confidence and number competence.

Parent Support

Since most parents place a high priority on attention to number-related topics, they will appreciate the inviting and repeatable activities in this book. Because most parents are willing to share the responsibility for repeated, short periods of practice, the following items are designed to promote parent involvement: *Family Letter* (p. 8), and *Questions Sampler* (p. 9). The first home packet might include the *Family Letter,* the *Questions Sampler,* and *Unaligned Products* (pp. 60–61). Since some students will benefit by reviewing the basic facts, *Line Ups Score* (p. 20–21) and *Multiplication Facts Made Easy* (p. 10) might be considered for a home packet. This packet should include materials for making three Number Cubes, each labeled with digits 1–6 (p. 153).

Students enjoy and benefit from repeated use of *Valuable Equations* (pp. 15–16) and *Finding Products* (p. 55). *Valuable Equations* require one set of Digit Squares, without the use of zero. Advise students and their families to keep the Number Cubes and Digit Squares in a safe place for repeat use throughout the school year. These Sponges lend themselves to home packets as well. Often popular Sponges can easily be converted into Games. Students can be challenged to assist with their creation. The advantage of Sponges, unlike Games, is that many of them can be experienced while a monitoring family member prepares dinner, packs lunches, or attends to other household tasks.

Family Letter

Dear Family,

To be prepared to work in the 21st century, all students need to be confident and competent in mathematics. Today the working world requires understanding of all areas of mathematics including statistics, logic, geometry, and probability. To be successful in these areas, students must maintain their basic facts and be able to compute. It is important that we be more efficient and effective in the time we devote to number-related topics. You can help your child in this area.

Throughout the school year, our mathematics program will focus on enhancing your child's understanding of number concepts. However, students must devote time at school and at home to practice and to improve these skills. Periodically, I will send home activities and related activity sheets that will build number sense and provide much needed practice. These games and activities have been carefully selected to engage your child in practicing more math skills than are usually answered on a typical page of drill.

By using the enclosed Questions Sampler during homework sessions, you will be able to assist your child without revealing the answers. The questions are categorized to help you select the most appropriate questions for your situation. If your child is having difficulty getting started with a homework assignment, try one of the questions in the first section. If your child gets stuck while completing a task, ask one of the questions from the second section. Try asking one of the questions from the third section to have your child clarify his or her mathematical thinking or to reflect on reasonableness of the results.

Good questions will help your child make sense of the mathematics, build confidence, and improve mathematical thinking and communication. I recommend posting the *Questions Sampler* in a convenient place so that you can refer to it often while helping your child with homework.

Your participation in this crucial area is most welcome.

Sincerely,

Questions Sampler

Getting Started

How might you begin?

What do you know now?

What do you need to find out?

While Working

How can you organize your information?

How can you make a drawing (model) to explain your thinking?

What approach (strategy) are you developing to solve this?

What other possibilities are there?

What would happen if . . . ?

What do you need to do next?

What assumptions are you making?

What patterns do you see? . . . What relationships?

What prediction can you make?

Why did you . . . ?

Checking Your Solutions

How did you arrive at your answer?

Why do you think your solution is reasonable?

What did you try that didn't work?

How can you convince me your solution makes sense?

Expanding the Response

(To help clarify your child's thinking, avoid stopping when you hear the "right" answer and avoid correcting the "wrong" answer. Instead, respond with one of the following.)

Why do you think that?

Tell me more.

In what other way might you do that? What other possibilities are there?

How can you convince me?

Multiplication Facts Made Easy

Since the tables go from 1 to 10, it appears there are 100 "facts" to memorize. However...

If you eliminate the easy 1s and 10s, that's **36 fewer facts.**

	1	2	3	4	5	6	7	8	9	10
1	~~1~~	~~2~~	~~3~~	~~4~~	~~5~~	~~6~~	~~7~~	~~8~~	~~9~~	~~10~~
2	~~2~~	4	6	8	10	12	14	16	18	~~20~~
3	~~3~~	6	9	12	15	18	21	24	27	~~30~~
4	~~4~~	8	12	16	20	24	28	32	36	~~40~~
5	~~5~~	10	15	20	25	30	35	40	45	~~50~~
6	~~6~~	12	18	24	30	36	42	48	54	~~60~~
7	~~7~~	14	21	28	35	42	49	56	63	~~70~~
8	~~8~~	16	24	32	40	48	56	64	72	~~80~~
9	~~9~~	18	27	36	45	54	63	72	81	~~90~~
10	~~10~~	~~20~~	~~30~~	~~40~~	~~50~~	~~60~~	~~70~~	~~80~~	~~90~~	~~100~~

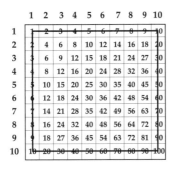

If you eliminate the 5s and 2s, which also seem easy, you will have **28 less.**

	1	2	3	4	5	6	7	8	9	10
1										
2		~~4~~	~~6~~	~~8~~	~~10~~	~~12~~	~~14~~	~~16~~	~~18~~	
3		~~6~~	9	12	~~15~~	18	21	24	27	
4		~~8~~	12	16	~~20~~	24	28	32	36	
5		~~10~~	~~15~~	~~20~~	~~25~~	~~30~~	~~35~~	~~40~~	~~45~~	
6		~~12~~	18	24	~~30~~	36	42	48	54	
7		~~14~~	21	28	~~35~~	42	49	56	63	
8		~~16~~	24	32	~~40~~	48	56	64	72	
9		~~18~~	27	36	~~45~~	54	63	72	81	
10										

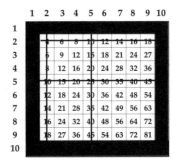

If the **squares,** such as three 3s and six 6s, seem easy to remember, you can **eliminate another 6 facts.**

	1	2	3	4	5	6	7	8	9	10
1										
2										
3			~~9~~	12		18	21	24	27	
4			12	~~16~~		24	28	32	36	
5										
6			18	24		~~36~~	42	48	54	
7			21	28		42	~~49~~	56	63	
8			24	32		48	56	~~64~~	72	
9			27	36		54	63	72	~~81~~	
10										

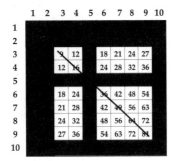

That leaves only 30 facts to memorize. However, we still have duplicates like 3 x 4 and 4 x 3; so we can cut 30 in half, **eliminating 15 more**.

	1	2	3	4	5	6	7	8	9	10
1										
2										
3				~~12~~		~~18~~	~~21~~	~~24~~	~~27~~	
4			12			~~24~~	~~28~~	~~32~~	~~36~~	
5										
6			18	24			~~42~~	~~48~~	~~54~~	
7			21	28		42		~~56~~	~~63~~	
8			24	32		48	56		~~72~~	
9			27	36		54	63	72		
10										

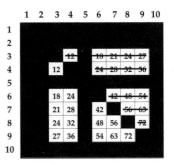

How many **"difficult" facts** are left to memorize? **100 – 36 – 28 – 6 – 15 = only ???**

4 × 3	**6 × 3**	**7 × 3**	**8 × 3**	**9 × 3**
6 × 4	**7 × 4**	**8 × 4**	**9 × 4**	**7 × 6**
8 × 6	**9 × 6**	**8 × 7**	**9 × 7**	**9 × 8**

Matrix–5/6 Games and Sponges

Type	Title	Topic	Page	Materials	Class	Groups	Pairs
G	**3-2-1 Zero**	Multiplication & Division	64	Digit Cards, Scratch Paper, Form p. 65		✓	✓
S	**Classmate Bingo**	Mental Computation	32	Markers, Forms pp. 33–34	✓	✓	
S	**Common Factors Match**	Common Factors	135	Form p. 136	✓	✓	
S	**Decimal Line-ups**	Multiplying Decimals	97	Markers, Calculator, Transparent Form p. 98	✓	✓	
G	**Divisible Products**	Multiplication & Division	62	Number Cubes (1–6), Number Cubes (4–9), Form p. 63			✓
G	**Figuring Percents**	Computing Percents	124	Markers, Calculator, Gameboards pp. 125–126			✓
S	**Finding Decimal Differences**	Subtracting Decimals	96		✓	✓	
S	**Finding Products**	Mental Multiplication & Division	55		✓	✓	
G	**Fraction Arrangements**	Computing Fractions	85	Digit Cards, Forms pp. 86–87			✓
S	**Going for More**	Computing Decimals	99	Transparent Digit Squares, Form p. 100	✓	✓	
G	**Greatest Common Factor Bingo**	Greatest Common Factors	142	Markers, Paper Clips & Pencil, Gameboard p. 143			✓
G	**High-Low**	Mental Computation	42	Number Cubes (1–6), Form p. 43			✓
S	**Identifying Values**	Visually Representing Relationships	137	Prepared Problems	✓	✓	
G	**Line Ups Score**	All Facts	20	Number Cubes (1–6), Markers, Gameboard p. 21			✓
G	**Matches Count**	All Facts	22	Form p. 23			✓
G	**Neighbors Count**	All Decimal Operations	108	Number Cubes (1–6) & (.1–.6), Markers, Scratch Paper, Gameboard p. 109			✓
S	**Operation Fill**	Mental Computation	35	5 × 8 cards, Transparent Form, p. 36	✓	✓	
G	**Ordered Fractions**	Comparing Fractions	83	Digit Cards, Form p. 84			✓
S	**Ordering Percents, Decimals, & Fractions**	Comparing Percents, Decimals, & Fractions	118	Transparent Digit Squares	✓	✓	
S	**Placing Percents**	Comparing Percents, Decimals, & Fractions	114	Digit Squares, Scratch Paper, Form p. 115	✓	✓	
S	**Product/Quotient Targets**	Multiplication & Division	56	Transparent Digit Squares, Container, Scratch Paper	✓	✓	
G	**Qualifying Equations**	Mental Computation	47	Number Cubes (1–6), Scratch Paper, Form p. 48			✓
G	**Reach the Peak**	Mental Computation	44	Number Cubes (4–9), Markers, Gameboards pp. 45-46			✓
S	**Remainders Count**	Division with Remainders	54	Digit Squares including Transparent Set, Scratch Paper	✓	✓	

S = Sponge G = Games

Blackline Masters: Digit Cards p. 150, Digit Squares p. 151, Number Cubes (1–6, blank) p. 153, Number Cubes (4–9, .1–.6) p. 154

Matrix–5/6 Games and Sponges

Type	Title	Topic	Page	Materials	Class	Groups	Pairs
G	**Rolling Decimals**	Converting Fractions to Decimals	104	Number Cubes (1–6), Form p. 105			✓
G	**Seek and Cover**	Solving for an Unknown	145	Number Cube (2,3,4,5,6, choose), Markers, Gameboard p. 146			✓
S	**Seeking 100**	Mental Computation	37	Markers, Opaque Covers, Transparent Form p. 38	✓	✓	
S	**Seeking Fractions**	Computing Fractions	76	Digit Squares, Form p. 77	✓	✓	
G	**Sixty**	Number Theory	144	Digit Cards, Scratch Paper			✓
S	**Spinning Fractions**	Computing & Ordering Fractions	72	Transparent Digit Squares, Paper Clip & Pencil, Scratch Paper, Form p. 73	✓	✓	
G	**Take the Discount**	Computing Discounted Prices	127	Markers, Paper Clip & Pencil, Scratch Paper, Gameboard p. 128		✓	
S	**Target 100%**	Adding Fractions & Percents	116	Markers, Form p. 117	✓	✓	
G	**Target 300%**	Adding Fractions, Decimals, & Percents	122	Markers, Gameboard p. 123			✓
S	**Target Fractions**	Computing Fractions	74	Digit Squares including Transparent Set, Paper Clip & Pencil, Form p. 75	✓	✓	
G	**Ten Tallies Win**	Reducing & Comparing Fractions	81	Number Cubes (1–6), Scratch Paper, Form p. 82			✓
S	**Today's Number**	All Facts	14	Scratch Paper	✓	✓	
G	**Totaling One**	Adding Decimals	106	Markers, Gameboard p. 107			✓
G	**Triangles Win**	All Facts	24	Number Cubes (1–6), Number Cube (7–12), Markers, Gameboard p. 25			✓
G	**Unaligned Products**	Mental Multiplication & Division	60	Markers, Gameboards p. 61			✓
S	**Valuable Equations**	All Facts	15	Transparent Digit Squares, Form p. 16	✓	✓	
S	**What's My Number?**	Symbolic Representations	138	Prepared Clues	✓	✓	
S	**Where's the Point?**	Decimal Relationships	94	Transparent Form p. 95	✓	✓	
S	**Which Doesn't Belong?**	Number Theory Concepts	134	Prepared Problems	✓	✓	

S = Sponge G = Game

Blackline Masters: Digit Cards p. 150, Digit Squares p. 151, Number Cubes (1–6, blank) p. 153, Number Cubes (4–9, .1–.6) p. 154

Mixed Facts

Assumptions The addition/subtraction and multiplication/division facts have previously been taught and reviewed, emphasizing understanding and number relationships. Concrete objects and visual models, such as counters and grids, have been used extensively.

Section Overview and Suggestions

Sponges

Today's Number p. 14

Valuable Equations pp. 15–16

These repeatable, whole-class or small-group warm-ups reinforce all facts and involve minimum preparation. Repeated use of both sponges will ensure greater success with the Games and Independent Activities in this section, especially the use of Valuable Equations prior to the Valuable Equations Practice activity.

Skill Checks

Quick Checks 1–6 pp. 17–19

The Skill Checks provide a way for parents, students, and you to see students' improvement with the basic facts. Copies may be cut in half so that each check may be used at a different time. Be sure to have all students respond to the STOP, number sense task, before solving the ten problems.

Games

Line Ups Score pp. 20–21

Matches Count pp. 22–23

Triangles Win pp. 24–26

These open-ended and repeatable Games actively involve students in practicing multiple facts as they enhance their mental computation abilities. *Triangles Win* includes a second gameboard to provide practice of easier facts. Frequent and repeated use of these games will increase students' confidence and competence in the "basics."

Independent Activities

Valuable Equations Practice p. 27

Joining Neighbors pp. 28–29

Matches Count Practice p. 30

These activities require students to independently practice many facts while solving engaging tasks. As students seek equations that qualify, they practice additional facts. *Valuable Equations Practice* is intended for repeated use, providing long-term practice. Note: the first format requires insertion of pre-selected numbers. The open-ended formats of *Joining Neighbors* and *Matches Count Practice* can be duplicated with new digit choices and grids to provide ongoing facts practice.

Today's Number

Topic: All Facts (+, –, ×, ÷)

Object: Create expressions that generate high scores.

Groups: Whole class or small group

Materials

- chalkboard or overhead projector
- scratch paper
- Digit Squares, p. 151 (*optional*)

Tips *Increase the number of possible expressions and the chances of students' success by allowing the displayed digits to be used more than once in an expression. Some students will benefit by manipulating Digit Squares or small pieces of paper.*

Directions

1. The leader displays five single-digit numbers and announces the day of the month as the target number.
2. The students are challenged to use any of the displayed numbers and any operations to create different expressions that equal today's target number. Displayed digits may only be used once in each expression.
3. Students earn one point for each different operation used in each generated expression. If all five digits are used, an additional point is awarded.
   Example: 4, 2, 9, 6, 3 are displayed and 15 is the target number.
   Expressions equaling 15: $(6 \times 2) + (4 \times 3) - 9$ (*4 points*)
   $6 + 4 + 3 + 2$ (*1 point*)
   $(4 \times 9) \div 2 - 3$ (*3 points*)
4. After students have adequate time to create and score their expressions, they exchange papers with a partner to have expressions and scores checked.
5. If the work is correct, the partner initials and returns the paper. When an error is discovered, the student who wrote the expression is allowed to create another one that works.
6. Expressions are shared with the entire class, showcasing a variety of possibilities.
7. If time allows, a student with the highest score announces a new target number and repeats the process. (It's more interesting when the announced number has some relevance to the class or personal relevance for the student.)

Making Connections

Promote reflection and make mathematical connections by asking:

- What approach worked to easily find varied solutions?
- How could the scoring be changed?

Valuable Equations

Topic: All Facts (+, –, ×, ÷)

Object: Create equations that generate high scores.

Groups: Whole class or small group

Materials

- *Valuable Equations* recording sheet for each student, p. 16
- transparent Digit Squares (1–9, zero removed), p. 151

Tips *Some teachers introduce this activity by displaying all four digits before students assign values and insert operation signs. This sponge can easily become a two player game.*

Directions

1. The leader draws and displays one Digit Square.
2. Students individually record the displayed amount in one of the four shapes of their Assigned Values key on their recording sheets. Students record that same value in each matching shape on the sheet.
3. Another Digit Square is drawn and displayed. Each student assigns this value to a new shape on her or his recording sheet.
4. This process is repeated until four Digit Squares are displayed and all shapes are filled by corresponding values.
5. Students independently determine what operations to place in each dotted parallelogram. Each student must use each operation at least once and is limited to no more than three uses of any operation sign. No expression can equal more than 120.
6. Each student finds the values of his or her six expressions. Students total the six values to determine their individual scores. The highest score wins.
7. After partners check each other's solutions, scores, and results to be certain all conditions listed in direction 5 have been followed, the equations and scores are shared with the entire class. Students usually do much better, when allowed to immediately play an additional round.

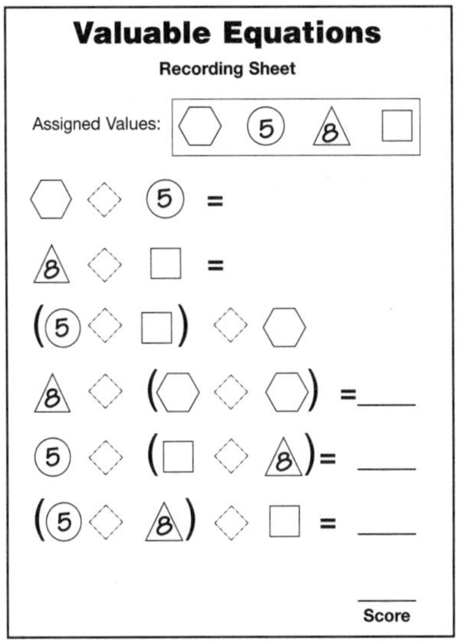

Making Connections

Promote reflection and make mathematical connections by asking:

- What placement patterns generated answers with high scores?
- What did you consider when placing division signs?...multiplication signs?

Valuable Equations

Recording Sheet

Assigned Values: 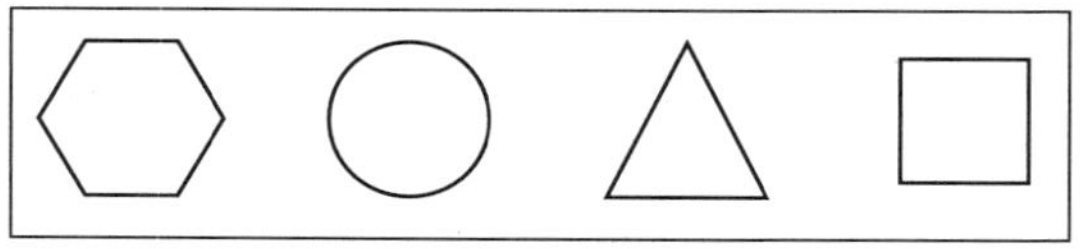

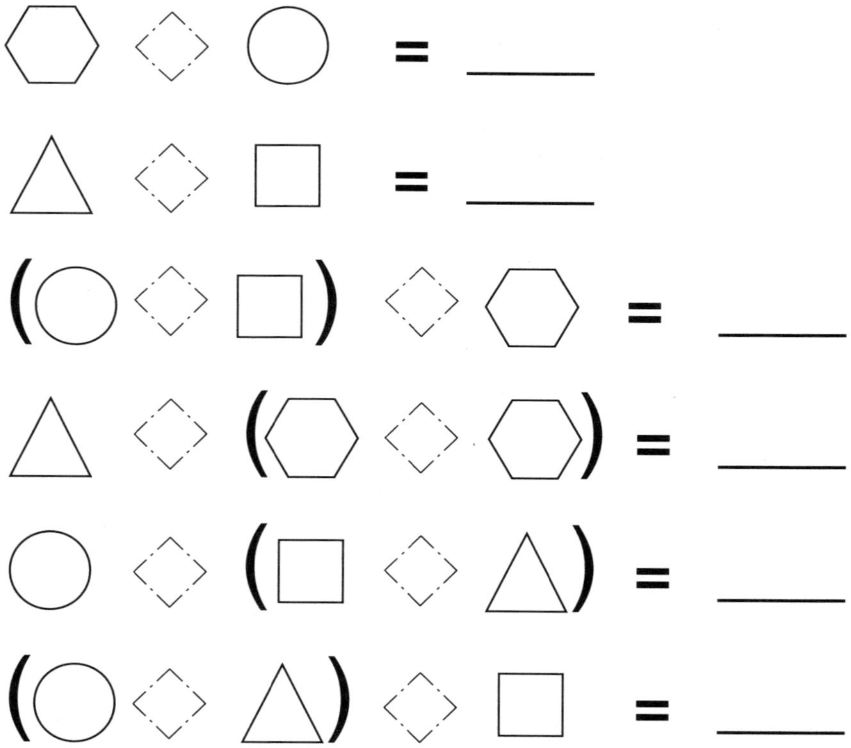

Score

Valuable Equations

Recording Sheet

Assigned Values:

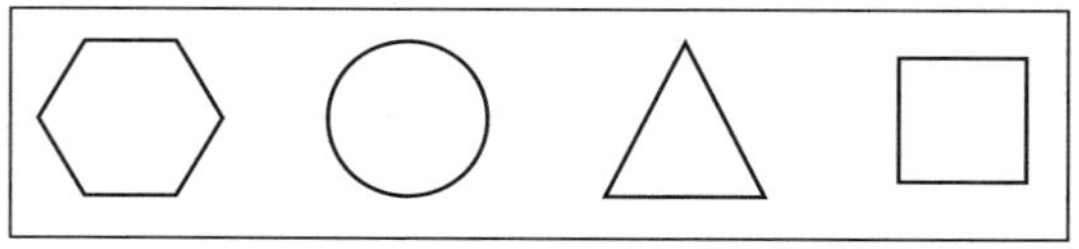

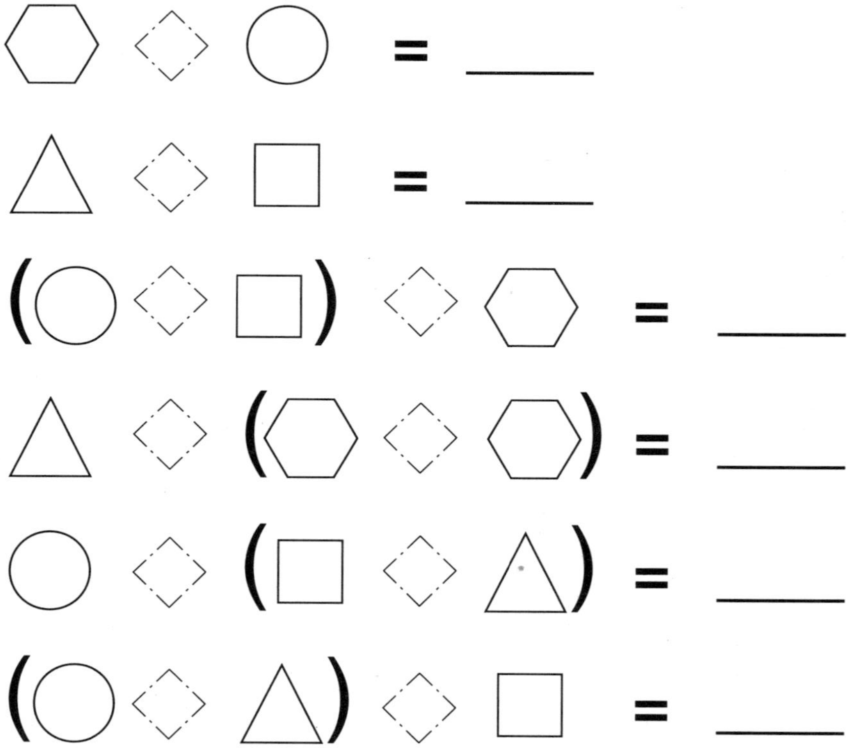

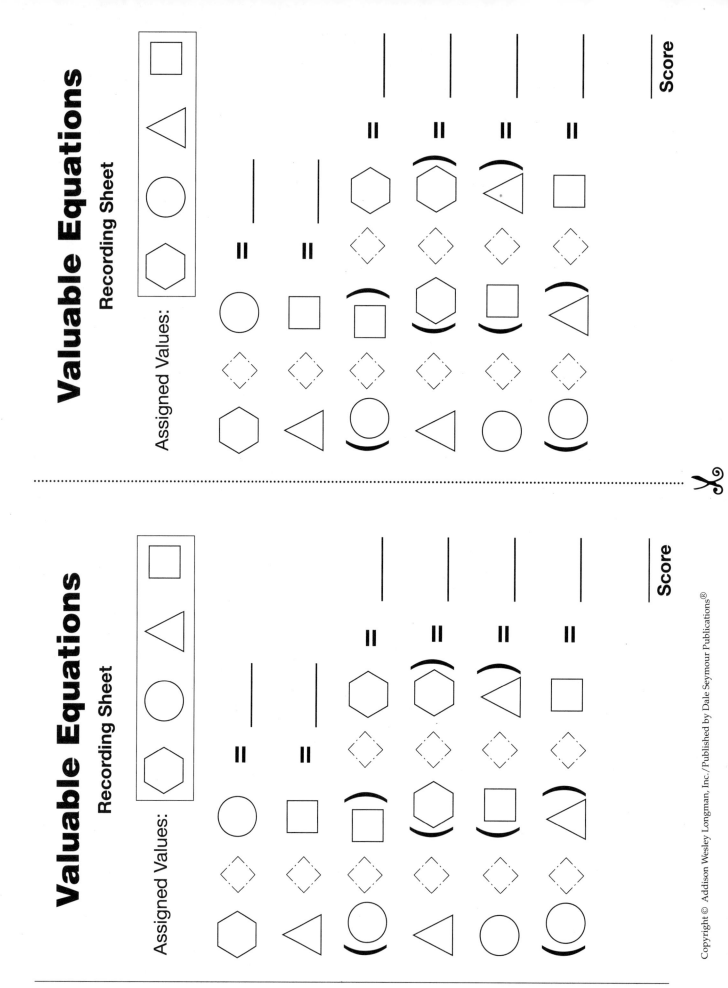

Score

Date ____________ Name ____________

Quick Checks 1

STOP Don't start yet. Star problems in the top row that may have even answers.

1. $(4 \times 7) + (30 \div 5) =$ ____ **2.** $(36 \div 9) \times (63 \div 7) =$ ____ **3.** $70 - (9 \times 5) =$ ____

Complete the following equations.

4. Use 4, 8 and 9.

$(\square \times \square) - \square = 23$

5. Use 3, 4 and 8.

$\square \times \square \div \square = 6$

6. Use 2, 3, 6 and 8.

$(\square + \square) \times (\square \div \square) = 36$

7. Use 2, 3, 4 and 9.

$(\square \div \square) \times (\square - \square) = 12$

Add operation signs and parentheses.

8. 6 ◯ 4 ◯ 3 = 8

9. 9 ◯ 7 ◯ 4 = 4

10. 4 ◯ 9 ◯ 7 ◯ 4 = 12

Go On Using at least two operations, write three equations that equal 23.

Date ____________ Name ____________

Quick Checks 2

STOP Don't start yet. Star a problem in the top row that may have the largest answer.

1. $(6 \times 7) - (32 \div 8) =$ ____ **2.** $(56 \div 7) \times (42 \div 7) =$ ____ **3.** $80 - (6 \times 8) =$ ____

Complete the following equations.

4. Use 6, 7 and 8.

$(\square \times \square) - \square = 34$

5. Use 2, 6 and 9.

$\square \times \square \div \square = 3$

6. Use 3, 4, 6 and 9.

$(\square + \square) \times (\square \div \square) = 30$

7. Use 1, 3, 6 and 9.

$(\square \div \square) \times (\square - \square) = 16$

Add operation signs and parentheses.

8. 9 ◯ 3 ◯ 6 = 9

9. 9 ◯ 7 ◯ 8 = 2

10. 5 ◯ 4 ◯ 12 ◯ 4 = 27

Go On Using at least two operations, write three equations that equal 19.

Date ____________ Name ____________

Quick Checks 3

STOP Don't start yet. Star problems in the top row that may have odd answers.

1. $(3 \times 8) - (42 \div 6) =$ ____ **2.** $(45 \div 9) \times (56 \div 8) =$ ____ **3.** $60 - (6 \times 6) =$ ____

Complete the following equations.

4. Use 5, 6 and 8.

$(\square \times \square) - \square = 22$

5. Use 3, 4 and 6.

$\square \times \square \div \square = 8$

6. Use 2, 4, 7 and 8.

$(\square + \square) \times (\square \div \square) = 18$

7. Use 1, 2, 6 and 8.

$(\square \div \square) \times (\square - \square) = 21$

Add operation signs and parentheses.

8. 8 ◯ 6 ◯ 4 = 12

9. 4 ◯ 3 ◯ 3 = 21

10. 7 ◯ 5 ◯ 9 ◯ 3 = 4

Go On Write as many equations as you can to equal today's day of the month. Describe your approach.

Date ____________ Name ____________

Quick Checks 4

STOP Don't start yet. Star a problem in the top row that may have the smallest answer.

1. $(7 \times 7) + (25 \div 5) =$ ____ **2.** $(54 \div 6) \times (49 \div 7) =$ ____ **3.** $80 - (8 \times 7) =$ ____

Complete the following equations.

4. Use 3, 7 and 9.

$(\square \times \square) - \square = 20$

5. Use 2, 4 and 6.

$\square \times \square \div \square = 3$

6. Use 2, 3, 4 and 6.

$(\square + \square) \times (\square \div \square) = 21$

7. Use 1, 3, 8 and 9.

$(\square \div \square) \times (\square - \square) = 21$

Add operation signs and parentheses.

8. 8 ◯ 3 ◯ 6 = 4

9. 5 ◯ 9 ◯ 2 = 7

10. 6 ◯ 3 ◯ 4 ◯ 7 = 29

Go On Using at least two operations, write three equations that equal 37.

Date ______________ Name ______________________

Quick Checks 5

STOP Don't start yet. Star problems in the top row that may have answers with multiples of 5.

1. $(5 \times 9) + (35 \div 7) =$ ____ **2.** $(48 \div 8) \times (81 \div 9) =$ ____ **3.** $60 - (7 \times 5) =$ ____

Complete the following equations.

4. Use 7, 8 and 9.

$(\square \times \square) - \square = 47$

5. Use 3, 6 and 9.

$\square \times \square \div \square = 2$

6. Use 3, 4, 5 and 6.

$(\square + \square) \times (\square \div \square) = 18$

7. Use 2, 6, 8 and 9.

$(\square \div \square) \times (\square - \square) = 12$

Add operation signs and parentheses.

8. 6 ◯ 3 ◯ 4 = 22

9. 6 ◯ 12 ◯ 4 = 9

10. 9 ◯ 3 ◯ 8 ◯ 5 = 4

Go On What are the next numbers in this pattern? 4, 12, 6, 18, 12, ___, ___
Describe the pattern.

Date ______________ Name ______________________

Quick Checks 6

STOP Don't start yet. Star problems in the top row that may have answers with multiples of 3.

1. $(8 \times 6) - (36 \div 4) =$ ____ **2.** $(63 \div 9) \times (36 \div 6) =$ ____ **3.** $90 - (8 \times 9) =$ ____

Complete the following equations.

4. Use 4, 7 and 9.

$(\square \times \square) - \square = 29$

5. Use 2, 3 and 4.

$\square \times \square \div \square = 6$

6. Use 2, 4, 5 and 7.

$(\square + \square) \times (\square \div \square) = 24$

7. Use 3, 4, 7 and 9.

$(\square \div \square) \times (\square - \square) = 9$

Add operation signs and parentheses.

8. 9 ◯ 3 ◯ 5 = 22

9. 8 ◯ 7 ◯ 3 = 4

10. 8 ◯ 4 ◯ 6 ◯ 4 = 8

Go On Use any four numbers to complete the equations.

$(\square \times \square) - (\square \div \square) = 10$ $(\square \times \square) \div (\square + \square) = 4$

Line Ups Score

Topic: All Facts (+, –, ×, ÷)

Object: Score the most points by aligning similar markers.

Groups: 2 players or pair players

Tip *An interesting variation is to have both players use the same three rolled numbers.*

Materials for each group

- *Line Ups Score* gameboard, p. 21
- 3 number cubes (1–6)
- different kind of markers for each player (12 each)

Directions

10	8	20	2	15
5	3	12	9	4
11	6	1	7	13
7	10	5	24	6
9	2	3	4	8

1. The first player rolls the three number cubes to generate three numbers.
2. The first player uses all three numbers and any operations to equal one number on the gameboard. The player states the expression, receives agreement from opponent, and covers the resulting answer with her or his marker.
3. The second player rolls the three number cubes, states an expression that equals an uncovered number, receives agreement, and covers that number.
4. If a player is unable to create any expression to equal an uncovered number, the opposing player has an opportunity to create an expression and cover the resulting answer with her or his marker. If no qualifying expression is found, the player passes that turn.
5. Play continues until both players have all 24 markers placed.
6. Each player determines her or his final score by awarding points as follows:

   3 in a row (1 point)
   4 in a row (3 points)
   5 in a row (6 points)

   Alignment can be vertical, horizontal, or diagonal.
7. At the end of the game, each player explains how her or his specified score was achieved.

Making Connections

Promote reflection and make mathematical connections by asking:

- What strategies seem to generate higher totals?
- How might a scoring sheet be designed to help others understand player's total scores?

Line Ups Score

10	8	20	2	15
5	3	12	9	4
11	6	1	7	13
7	10	5	24	6
9	2	3	4	8

Matches Count

Topic: All Facts (+, –, ×, ÷)

Object: Score the most points.

Groups: 2 players or pair players

Materials for each group

- *Matches Count* recording sheet (for each player), p. 23

Directions

1. The first player selects and announces one number from the number bank.
2. Each player independently crosses out and records the announced number in one of her or his squares. Once a number is recorded it cannot be changed.
3. The second player selects and announces a new number which is independently recorded and crossed out on each player's recording sheet.
4. Players continue to take turns selecting, recording, and crossing out numbers until nine numbers have been announced.
5. Players independently use any operations to create eight expressions. Players should try to form expressions that have values matching at least one other expression. Division must result in whole number answers.
6. Only expressions with matching values are recorded on the recording lines.
7. The answers to the recorded matching expressions are totaled to determine each player's score.

Number Bank

~~1~~	2	~~3~~	~~4~~	~~5~~
~~6~~	~~7~~	~~8~~	~~9~~	10
11	12	13	14	~~15~~
16	17	18	19	20

Expressions with matching values:

1 x 4 x 9 = 36

15 + (7 x 3) = 36

(9 – 3) x 6 = 36

(4 + 7) – 8 = 3

(1 x 15) ÷ 5 = 3

Tips *Initially introduce this activity as a class game and discuss strategic placing of larger and smaller numbers. This game can effectively be introduced and played as a sponge with the class divided into two teams.*

Making Connections

Promote reflection and make mathematical connections by asking:

- When you play this again, what will you do differently?
- What strategy helped you make matches?

Matches Count

Number Bank

1	2	3	4	5
6	7	8	9	10
11	12	13	14	15
16	17	18	19	20

Expressions with matching values:

____

____

____

____

____

____

____

____

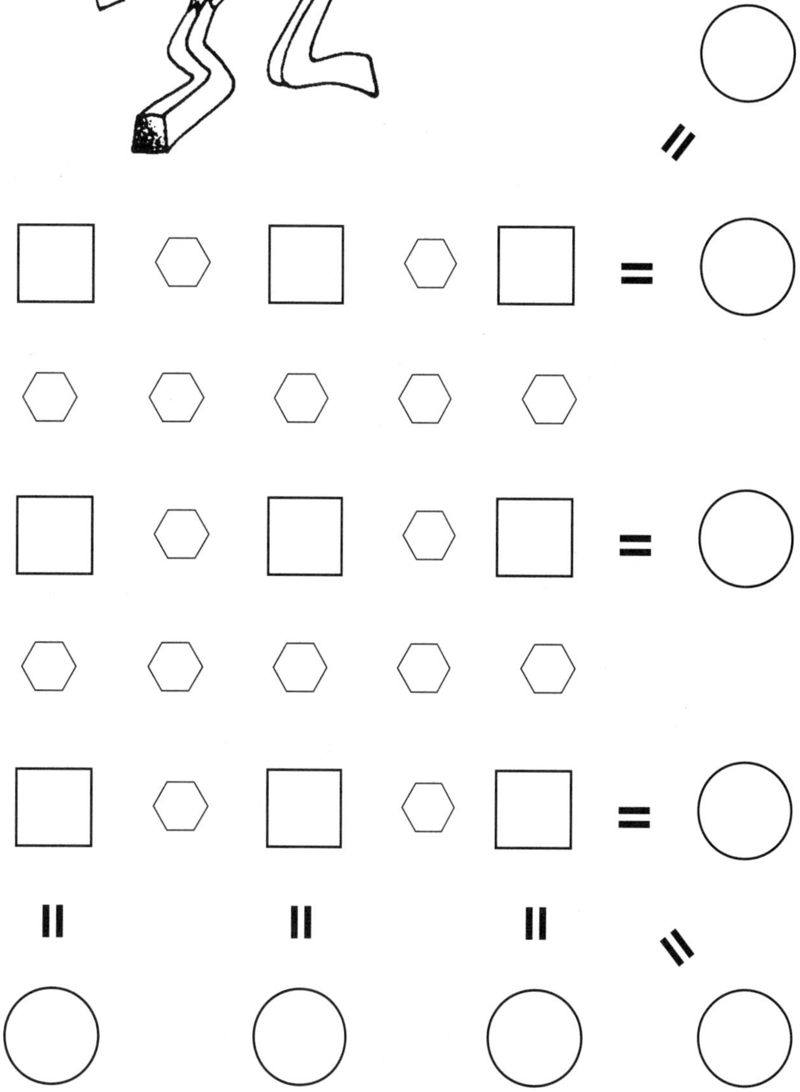

Triangles Win

Topic: All Facts (+, –, ×, ÷)

Object: Cover the vertices of any three equilateral triangles.

Groups: 2 players or pair players

Materials for each group

- *Triangles Win A* gameboard, p. 25
- 2 number cubes (1–6)
- special number cube (7–12), p.153
- different kind of markers for each player

Tip *If you prefer an easier version, use the Triangles Win B gameboard with three regular number cubes (1–6).*

Directions

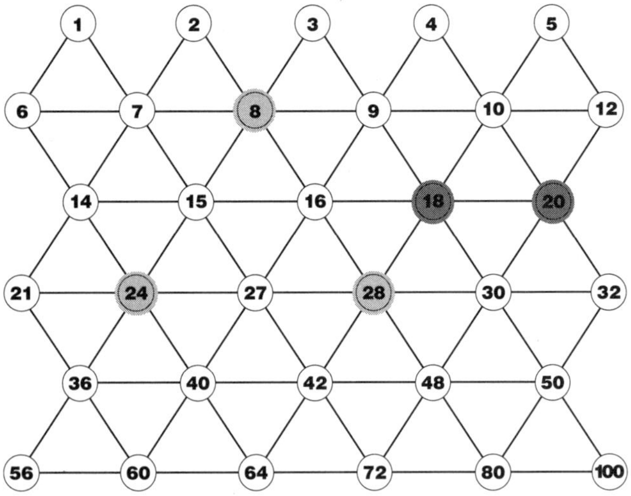

1. One player generates three numbers by rolling the number cubes.
2. Each player uses the three numbers and any operations to equal one number on the gameboard. When a desired solution is identified, the player announces and covers the resulting answer with her or his marker and states the complete equation. The opponent verifies the accuracy of the equation.
3. The other player covers a different uncovered number and states the equation. Only one marker is allowed on one space.
4. Players continue to roll, cover results, and state equations until one player positions her or his markers to form three equilateral triangles of any size on the gameboard.
5. If three rolled numbers generate only one available number, only the first player to announce the resulting number has a play.
6. The successful player identifies the three created triangles and wins.

Making Connections

Promote reflection and make mathematical connections by asking:

- Where are some good starting positions for this game?
- How did you usually use the highest number rolled? Please explain.

Triangles Win

Gameboard A

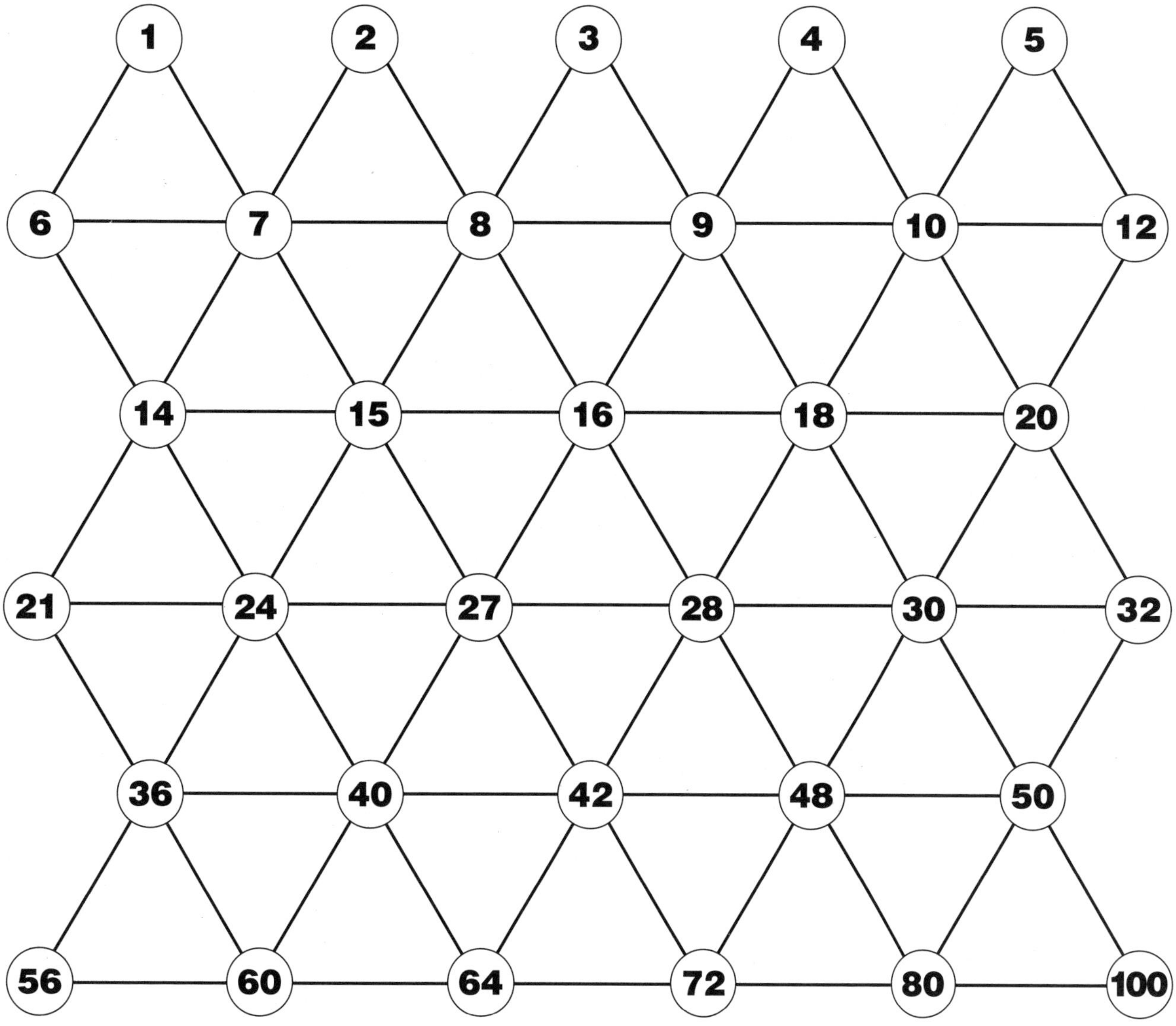

Triangles Win

Gameboard B

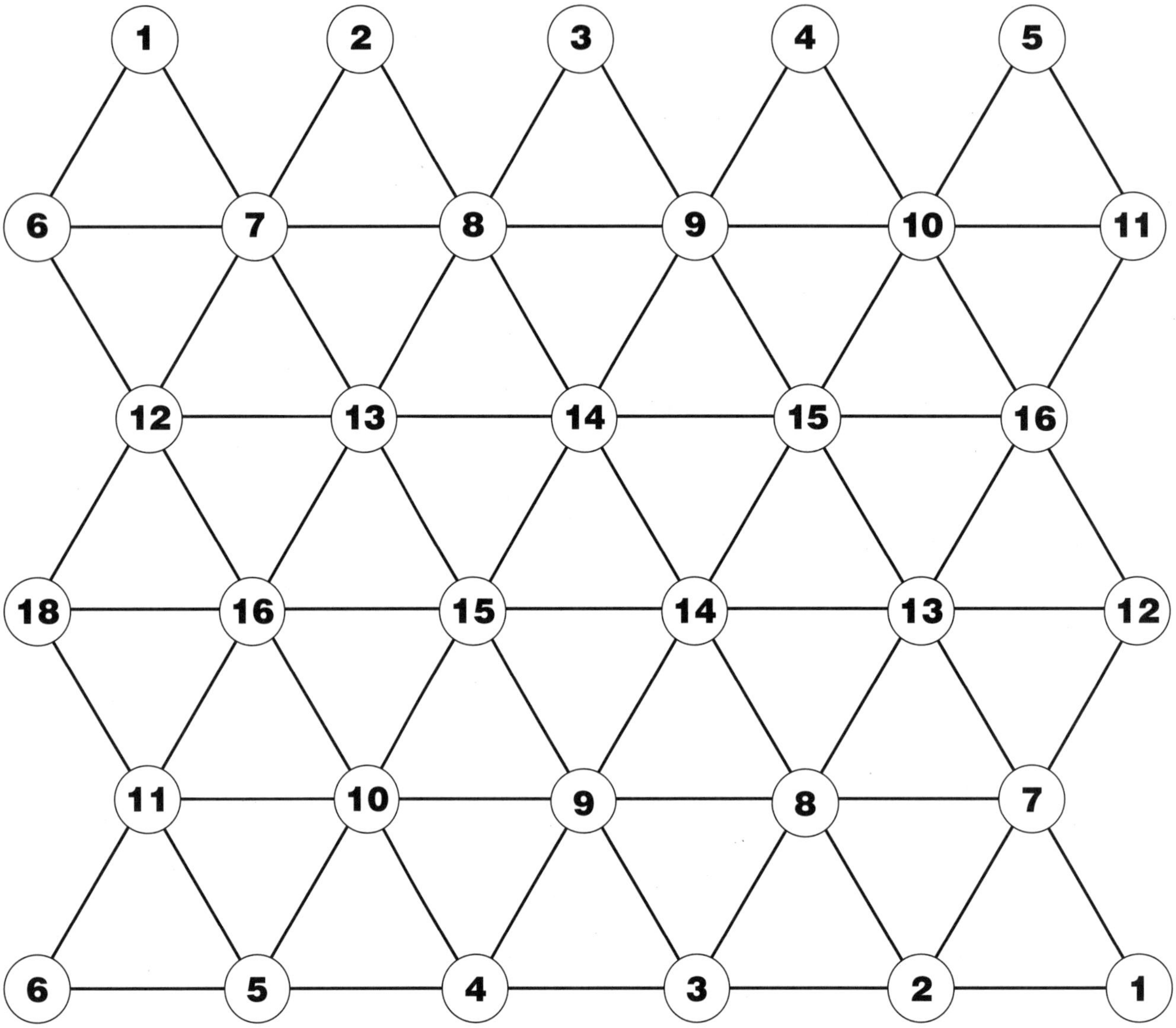

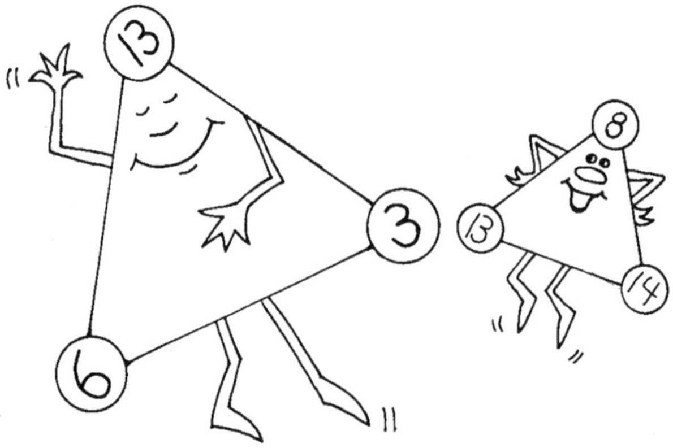

Date ________ Name ____________________________

Valuable Equations

Assign these numbers to each shape: ____ ____ ____ ____
Follow these rules as you solve each equation.
    Choose the operations to place in each dotted diamond.
    Use no more than 3 of any operation sign.
Total your answers to determine your score.
Try for a score over 150.

Assigned Values: ⬡ ○ △ □

⬡ ◇ ○ = ______

△ ◇ □ = ______

(○ ◇ □) ◇ ⬡ = ______

△ ◇ (⬡ ◇ ⬡) = ______

○ ◇ (□ ◇ △) = ______

(○ ◇ △) ◇ □ = ______

______
Score

Date ________ Name ____________________________

Valuable Equations

Assign 4 different single-digit numbers to the shapes.
Follow these rules as you solve each equation.
    Choose the operations to place in each dotted diamond.
    Use no more than 3 of any operation sign.
Total your answers to determine your score.
Try for a score over 150.

Assigned Values: ⬡ ○ △ □

⬡ ◇ ○ = ______

△ ◇ □ = ______

(○ ◇ □) ◇ ⬡ = ______

△ ◇ (⬡ ◇ ⬡) = ______

○ ◇ (□ ◇ △) = ______

(○ ◇ △) ◇ □ = ______

______
Score

Date ____________________ Name ______________________________

Joining Neighbors I

Create expressions by using at least three adjoining numbers and more than one operation to equal the indicated target numbers. Loop the used numbers on the smaller grids and record the equations below. Remember to use parentheses when necessary. An example has been done for you.

2	4	1	3	5
6	3	8	2	1
5	9	4	3	2
1	2	5	6	3
4	6	3	2	7

2	4	1	3	5
6	3	8	2	1
5	9	4	3	2
1	2	5	6	3
4	6	3	2	7

2

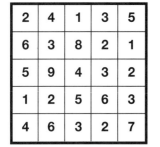

1. $5 \times 4 \div (8 + 2) = 2$
2. ____________________
3. ____________________

2	4	1	3	5
6	3	8	2	1
5	9	4	3	2
1	2	5	6	3
4	6	3	2	7

5

4. ____________________
5. ____________________
6. ____________________

2	4	1	3	5
6	3	8	2	1
5	9	4	3	2
1	2	5	6	3
4	6	3	2	7

4

7. ____________________
8. ____________________
9. ____________________

2	4	1	3	5
6	3	8	2	1
5	9	4	3	2
1	2	5	6	3
4	6	3	2	7

10

10. ____________________
11. ____________________
12. ____________________

Date ____________ Name ____________

Joining Neighbors II

Create expressions by using at least three adjoining numbers and more than one operation to equal the indicated target numbers. Loop the used numbers on the smaller grids and record the equations below. Remember to use parentheses when necessary.

4	9	3	7	2
1	6	2	8	5
7	5	4	3	8
2	9	6	1	4
5	8	3	2	6

4	9	3	7	2
1	6	2	8	5
7	5	4	3	8
2	9	6	1	4
5	8	3	2	6

10

1. ____________
2. ____________
3. ____________

4	9	3	7	2
1	6	2	8	5
7	5	4	3	8
2	9	6	1	4
5	8	3	2	6

3

4. ____________
5. ____________
6. ____________

4	9	3	7	2
1	6	2	8	5
7	5	4	3	8
2	9	6	1	4
5	8	3	2	6

7

7. ____________
8. ____________
9. ____________

4	9	3	7	2
1	6	2	8	5
7	5	4	3	8
2	9	6	1	4
5	8	3	2	6

12

10. ____________
11. ____________
12. ____________

Date ____________ Name ____________________

Matches Count Practice

Follow the directions for the *Matches Count* game (p. 22) to complete this activity.

1. Arrange the numbers

2 3 5 6 9 10 12 15 18

to produce a score over 50.

Divide at least once.

Record the equations with matching answers and total your score.

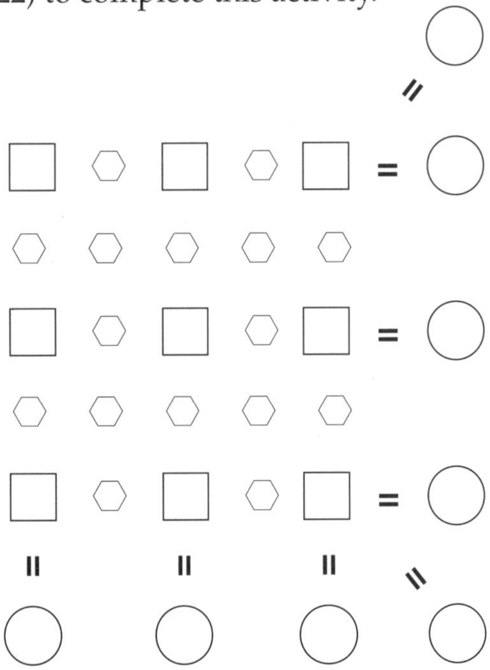

____________ ____________

____________ ____________

____________ ____________

____________ ____________

Score = _____

2. Arrange the numbers

1 2 3 4 5 6 8 10 16

to produce a score over 100.

Record the equations with matching answers and total your score.

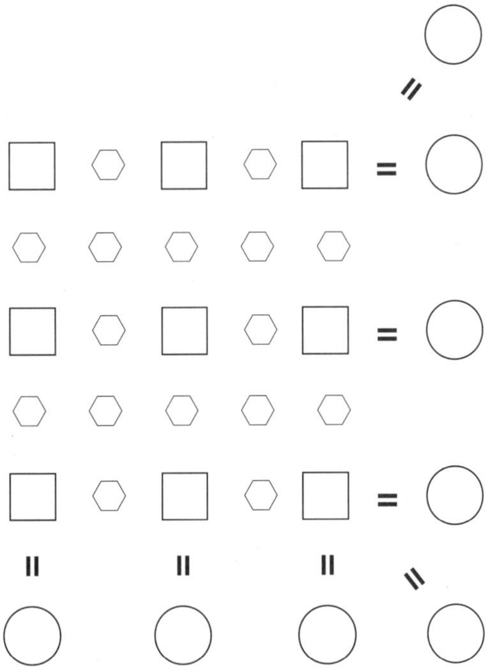

____________ ____________

____________ ____________

____________ ____________

____________ ____________

Score = _____

Mental Computation

Assumptions Computation operations have previously been taught and reviewed, emphasizing understanding and building operation sense. A variety of mental estimation and computation strategies have been explored and shared by students. Mental computation is promoted regularly and frequently.

Section Overview and Suggestions

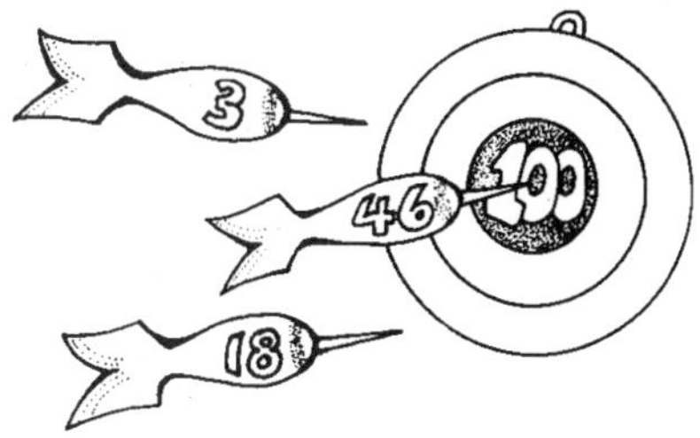

Sponges

Classmate Bingo pp. 32–34

Operation Fill pp. 35–36

Seeking 100 pp. 37–38

These whole-class or small-group warm-ups are repeatable. They reinforce all operations and require mental estimation and computation. Repeat use of these sponges will ensure greater success with the Games and Independent Activities in this section, especially the use of the challenging *Seeking 100* sponge.

Skill Checks

In Your Head 1–6 pp. 39–41

The Skill Checks provide a way for parents, students, and you to see students' improvement with mental computation. Copies may be cut in half so that each check may be used at a different time. Be sure to have all students respond to the STOP, number sense task, before solving the ten problems. These Skill Checks are unique because they attempt to promote a mental computation, rather than paper and pencil approach.

Games

High-Low pp. 42–43

Reach the Peak pp. 44–46

Qualifying Equations pp. 47–48

These very repeatable Games actively involve students in mental computation and build operation sense. With frequent use of these games at home, students will gain confidence and competence in their abilities to compute mentally. Both the regular and the expert versions of *Reach the Peak* are very challenging games for most students and require adequate playing time.

Independent Activities

Rearrange and Find pp. 49–50

Related Cross-Number Puzzle p. 51

Operation Fill Practice p. 52

These activities actively engage students in mental computation. Use of the *Operation Fill* sponge will ensure greater success with *Operation Fill Practice.* Both *Rearrange and Find* and *Operation Fill Practice* can be easily replicated with new numbers to provide further practice.

Classmate Bingo

Topic: Mental Computation

Object: Cover four numbers in a row.

Groups: Whole class or small group

Materials

- *Classmate Bingo* playing boards, p. 33
- *Classmate Bingo* clues, cut apart and placed in container, p. 34
- 10–15 markers for each student

Tips *Have students collaborate to create clues for additional rounds with this same range of numbers or create new Bingo playing boards and new clues for extended use of this warm-up.*

Directions

1. The leader asks students to create unique bingo playing boards by randomly recording the numbers 30 through 45 in their 16 cells.
2. The leader draws, reads, and displays one clue card.
3. Students individually solve the clue and cover the solution on their playing board.
4. Students agree on the correct response to the first clue before the leader goes on to a second clue.
5. The leader continues announcing and displaying clues as students solve and cover corresponding solutions.
6. When a student has four markers in a row horizontally, vertically, or diagonally, the student announces "Bingo."
7. If time allows, play continues until each student in the class has at least one Bingo.

Making Connections

Promote reflection and make mathematical connections by asking:

- Which clues were more difficult? Please explain.
- Which clue(s) requires more than one mental computation step? Please explain.

Classmate Bingo

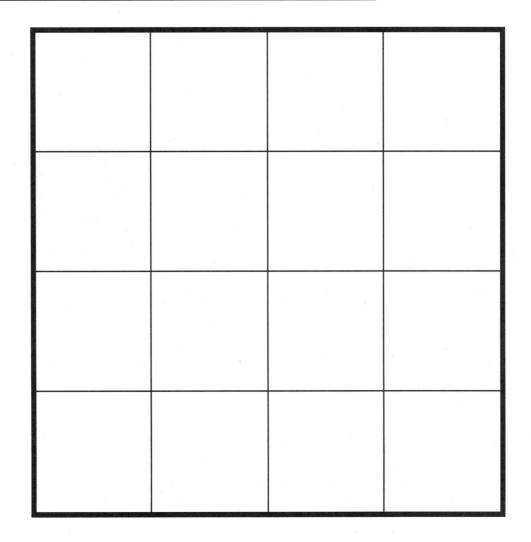

Classmate Bingo Clues

27,000 ÷ 900	even number between 34 and 42 that is not divisible by 4
(2 x 3 x 30) ÷ 9, plus the number of players on a football team	multiply this number by 3 and the result is 117
2 x 2 x 2 x 2 x 2	2400 ÷ 60
467 + ☐ = 500	the first prime number in the forties
the perimeter of a rectangle that is 9 by 8	168 ÷ 4
13 less than the number of inches in 4 feet	9 less than the number of weeks in a year
number of months in three years	100 less than a gross
19 x 3, less 20	number of degrees in half of a right angle

Operation Fill

Topic: Mental Computation

Object: Complete equations by inserting operation signs.

Groups: Whole class or small group

Materials

- transparency of *Operation Fill,* p. 36
- two 5-by-8 cards

Tip *If students are experiencing difficulty with this challenging sponge, initially identify the specific operation signs that need to be inserted.*

Directions

1. The leader displays only the first line of *Operation Fill,* and asks students to determine how to insert two operational signs in the number sequence to produce an accurate equation.
2. Without pencil and paper, students figure out how the sequence of displayed numbers can be combined with two different operation signs to equal 77.
3. After students have adequate time to determine the placement and identity of the operation signs, the completed equations are shared.
4. While the solved row is covered by another 5-by-8 card, the second row is displayed. Students are again asked to figure out without paper and pencil how to equal the displayed answer.

$$(2\quad 5 \times 4) - 2\quad 3 = 77$$

$$(1\quad 4\quad 2 - 6\quad 9) + 7 = 80$$

5. After these solutions are found and shared, students are informed that future rows may include two or three operation signs and that an operation may be repeated. (The last four sequences require three operation signs.)
6. After students have solved at least five equations, have them partner to create similar number-sequence puzzles for classmates to solve. If appropriate, require the insertion of at least one division or multiplication sign.
7. Student-authored number-sequence puzzles are recorded on the rows of lines on the bottom portion of the Operation Fill form.

Making Connections

Promote reflection and make mathematical connections by asking:

- What patterns did you see that helped you find solutions?
- What arrangements seemed to indicate a specific operation sign?

Operation Fill

2 5 4 2 3 = 77

1 4 2 6 9 7 = 80

4 8 6 1 5 = 120

7 5 4 2 9 = 329

3 1 2 1 5 8 2 = 77

4 8 3 4 3 6 = 100

7 1 4 5 9 5 = 45

5 1 2 4 2 9 4 = 400

1 3 5 1 3 8 6 = 135

Seeking 100

Tips Once students gain familiarity with this warm-up, they enjoy pairing up and playing it as a game. Allowing participants to use any placed markers or changing the target number to 150, provides a greater challenge.

Topic: Mental Computation

Object: Create five qualifying expressions.

Groups: Whole class or small group

Materials

- transparency of *Seeking 100,* p. 38
- two kinds of markers
- opaque covers to fit within cells

Directions

1. The leader divides the group into two teams.
2. Keeping in mind the goal of creating expressions that equal 100, members of the first team strategically select and cover one number with their marker.
3. The second team follows the same process.
4. Teams continue to alternate turns announcing and covering one number on each turn.
5. After three turns each, the teams attempt to use their three covered numbers to make an expression that equals 100.
6. If a team finds and announces an expression, the equation is recorded on the chalkboard and the markers covering those numbers are replaced by opaque covers. (The opaque covered numbers are not available for future equations.)
7. If a team cannot find an expression, the team covers a fourth number and repeats the process using at least three numbers in each expression.
8. This process continues until a team records five equations which include at least one division sign and no more than one equation with only addition signs. Both teams may finish at the same time.

TEAM B

$30 \times 3 + 10 = 100$

$36 \div 9 \times 25 = 100$

$15 \times 2 \times 5 - (44 + 6) = 100$

$17 \times 7 - 19 = 100$

Making Connections

Promote reflection and make mathematical connections by asking:

- How will you play this game differently next time?
- What explanation might be given for not using the uncovered numbers?

Seeking 100

1	2	3	4	5	6	7	8	9	10
11	12	13	14	15	16	17	18	19	20
21	22	23	24	25	26	27	28	29	30
31	32	33	34	35	36	37	38	39	40
41	42	43	44	45	46	47	48	49	50

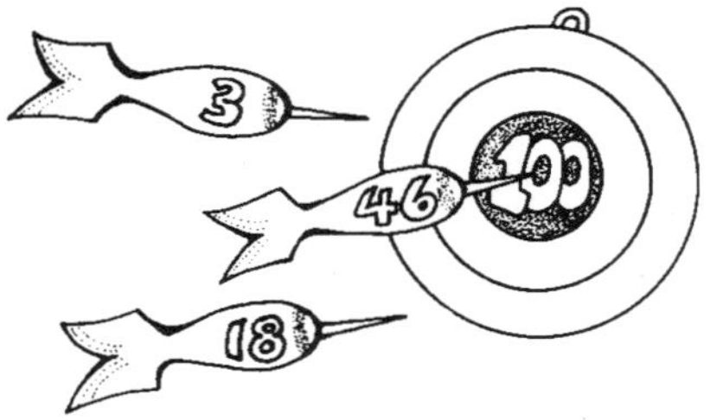

Date ____________ Name ____________________

In Your Head 1

STOP Don't start yet. Star 2 problems that may have odd answers.

Important: Try to compute each problem mentally, only recording your final answer.

1. 88 + 33 + 36 = ____ **2.** 362 + ____ = 500 **3.** 417 – 23 = ____

4. 3 × 7 × 40 = ____ **5.** 4 × 16 = ____ **6.** 54 × 3 = ____

7. 60 × 70 = ____ **8.** 29 × 30 = ____ **9.** 89 ÷ 7 = ____ **10.** 231 ÷ 3 = ____

Go On Use +, –, ×, or ÷ to make each equation true.
35 ◯ 480 ◯ 6 = 2800 250 ◯ (14 ◯ 9) = 124

Date ____________ Name ____________________

In Your Head 2

STOP Don't start yet. Star problems that may have answers between 100 and 300.

Important: Try to compute each problem mentally, only recording your final answer.

1. 97 + 52 + 23 = ____ **2.** 483 + ____ = 800 **3.** 734 – 98 = ____

4. 8 × 50 × 4 = ____ **5.** 18 × 8 = ____ **6.** 42 × 6 = ____

7. 400 × 50 = ____ **8.** 49 × 50 = ____ **9.** 120 ÷ 9 = ____ **10.** 522 ÷ 6 = ____

Go On Use the digits 3, 4, 6 and 8 to write a multiplication problem with the largest possible product. Describe your strategy.

Date ____________ Name ____________________

In Your Head 3

STOP Don't start yet. Star problems that may have answers that are multiples of 100.

Important: Try to compute each problem mentally, only recording your final answer.

1. 71 + 24 + 48 = ____

2. 374 + ____ = 600

3. 524 – 37 = ____

4. 30 × 5 × 4 = ____

5. 17 × 5 = ____

6. 36 × 4 = ____

7. 70 × 80 = ____

8. 19 × 40 = ____

9. 88 ÷ 6 = ____

10. 336 ÷ 4 = ____

Go On Use the digits 2, 4, 5, 6, and 7 to write a subtraction problem with the smallest possible difference. Describe your strategy.

Date ____________ Name ____________________

In Your Head 4

STOP Don't start yet. Star 2 problems that may have four-digit answers.

Important: Try to compute each problem mentally, only recording your final answer.

1. 25 + 89 + 62 = ____

2. 358 + ____ = 700

3. 848 – 76 = ____

4. 5 × 60 × 8 = ____

5. 9 × 19 = ____

6. 38 × 7 = ____

7. 80 × 600 = ____

8. 39 × 60 = ____

9. 107 ÷ 8 = ____

10. 532 ÷ 7 = ____

Go On Use +, –, ×, or ÷ to make each equation true.

675 ◯ (45 ◯ 30) = 45   540 ◯ (15 ◯ 9) = 90

Date ____________ Name ____________________

In Your Head 5

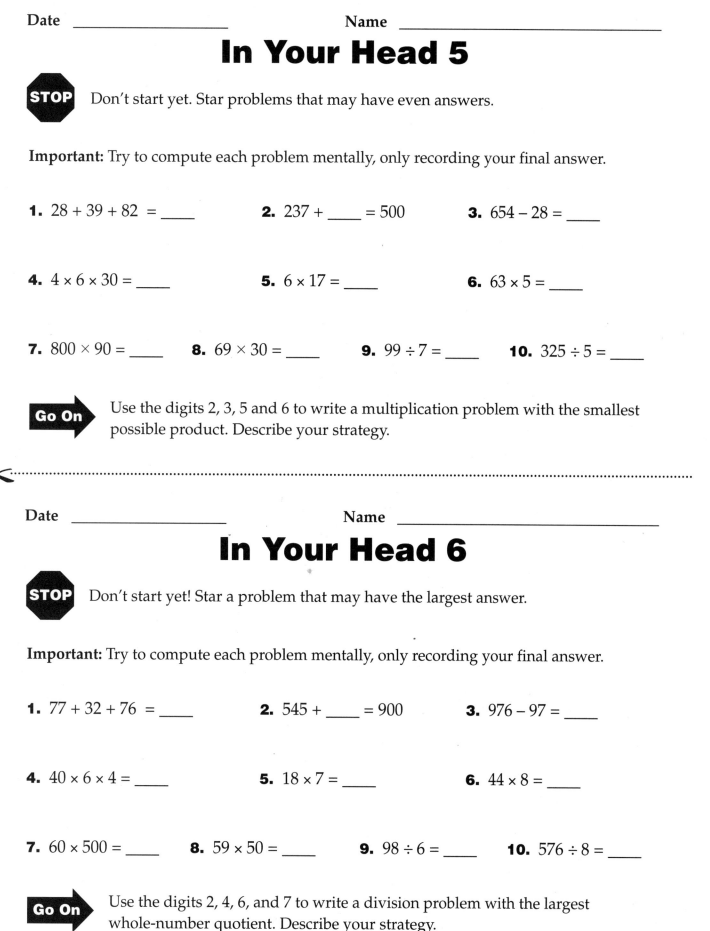

STOP Don't start yet. Star problems that may have even answers.

Important: Try to compute each problem mentally, only recording your final answer.

1. 28 + 39 + 82 = ____

2. 237 + ____ = 500

3. 654 – 28 = ____

4. 4 × 6 × 30 = ____

5. 6 × 17 = ____

6. 63 × 5 = ____

7. 800 × 90 = ____

8. 69 × 30 = ____

9. 99 ÷ 7 = ____

10. 325 ÷ 5 = ____

Go On Use the digits 2, 3, 5 and 6 to write a multiplication problem with the smallest possible product. Describe your strategy.

Date ____________ Name ____________________

In Your Head 6

STOP Don't start yet! Star a problem that may have the largest answer.

Important: Try to compute each problem mentally, only recording your final answer.

1. 77 + 32 + 76 = ____

2. 545 + ____ = 900

3. 976 – 97 = ____

4. 40 × 6 × 4 = ____

5. 18 × 7 = ____

6. 44 × 8 = ____

7. 60 × 500 = ____

8. 59 × 50 = ____

9. 98 ÷ 6 = ____

10. 576 ÷ 8 = ____

Go On Use the digits 2, 4, 6, and 7 to write a division problem with the largest whole-number quotient. Describe your strategy.

High-Low

Topic: Mental Computation

Object: Produce the greatest difference.

Groups: 2 players or pair players

Materials for each group

- *High-Low* recording sheet (for each player), p. 43
- 8 number cubes (1–6)

Tips *Students find it helpful to move the number cubes as they try to create high and low scores. Students discover parentheses can be strategically used to produce even higher and lower scores.*

Directions

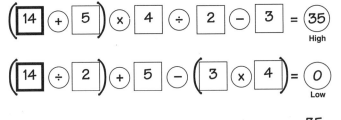

1. Each player rolls four number cubes and finds the sum of the generated numbers. The resulting sum is recorded in the two heavily outlined boxes in the first pair of equations on the recording sheet.
2. Each player uses the four generated numbers and each of the four operations to complete the expressions, seeking the highest and the lowest answers.
3. Players must honor order of operations and are encouraged to clarify their intentions by inserting parentheses. The division operation must produce a whole number quotient.
4. Once a player identifies a desired equation, he or she writes the equation on the recording sheet and places operation signs in the small circles.
5. After recording both equations, a player determines and records the difference between his or her high and low solutions.
6. Players exchange papers to have their results verified.
7. The winner is the player with correct equations and the greatest difference.
8. Players are encouraged to play additional rounds.

$$(14 + 5) \times 4 \div 2 - 3 = 35 \quad \text{High}$$

$$(14 \div 2) + 5 - (3 \times 4) = 0 \quad \text{Low}$$

Score = 35 (difference)

Making Connections

Promote reflection and make mathematical connections by asking:

- What operations did you frequently use with high numbers? ... with low numbers?
- What combinations seemed to produce the greatest differences?

High-Low Recording Sheet

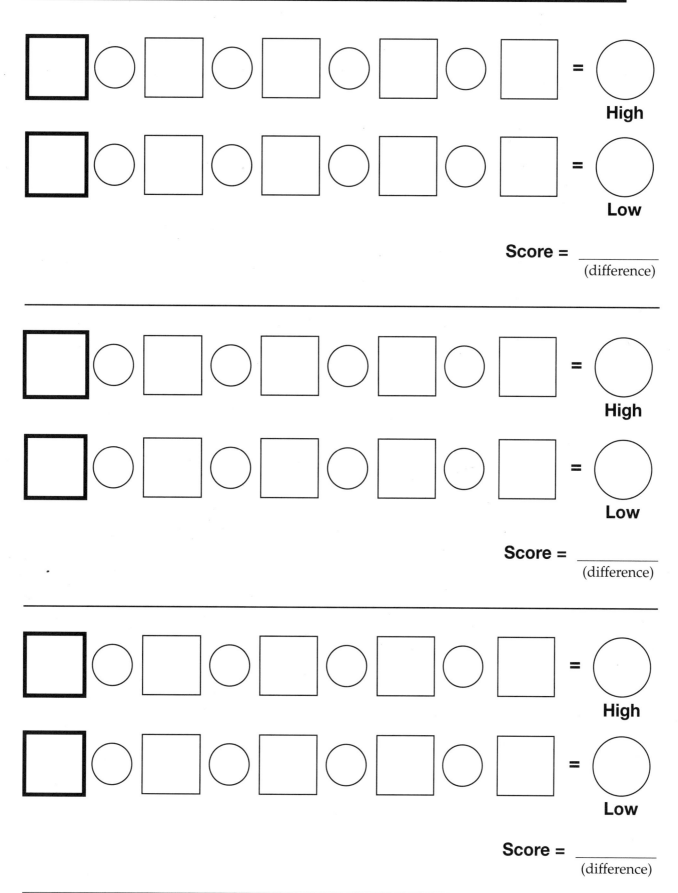

= High

= Low

Score = ______
(difference)

= High

= Low

Score = ______
(difference)

= High

= Low

Score = ______
(difference)

Reach the Peak

Topic: Mental Computation

Object: Travel from the bottom to the peak of the gameboard.

Groups: 2 players or pair players

Materials for each group

- *Reach the Peak* gameboard, p. 45
- special number cube (4–9), p. 154
- different transparent marker for each player

Tips If students become competent playing this game, challenge them to try the "Expert Level" version (p. 46). Different skills can easily be practiced using this format by writing new problems on the blank Reach the Peak *gameboard, p. 155, and a number cube of your choice.*

Directions

1. The first player rolls the number cube and then finds a problem in the bottom row that requires the rolled number to complete it. The player justifies the solution and covers the problem with his or her marker. During the game, players are allowed to record equation solutions that have been solved mentally.
2. The second player follows the same process and enters the gameboard on the bottom row. Players are allowed to share a cell.
3. The first player rolls again. If the player rolls a number that can complete a problem in an adjoining cell (defined as above, below, or to the side), the player must move to that cell.
4. If the player can't find an adjoining cell that requires the rolled number, the player passes and does not move for that turn.
5. If the other player identifies the rolled number as a match for an adjoining cell, the player who passed must move down one cell.
6. Players continue to alternate turns, rolling, solving problems, moving to adjoining cells, and justifying actions aloud.
7. The winner is the first player to reach the peak (top triangular cell).

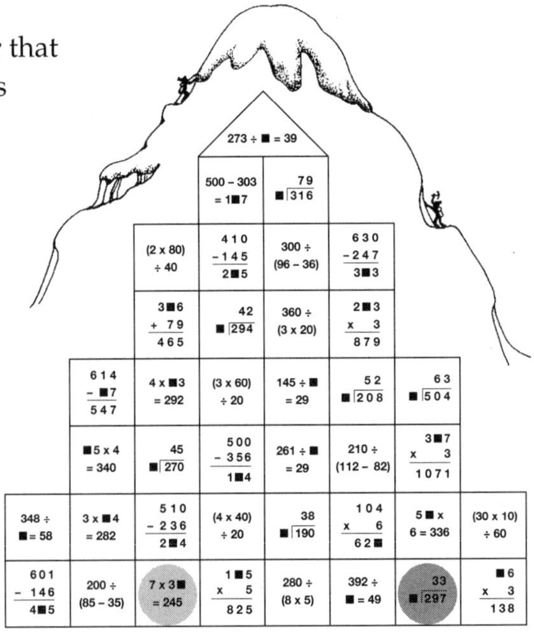

Making Connections

Promote reflection and make mathematical connections by asking:

- How could this game be improved?

Reach the Peak

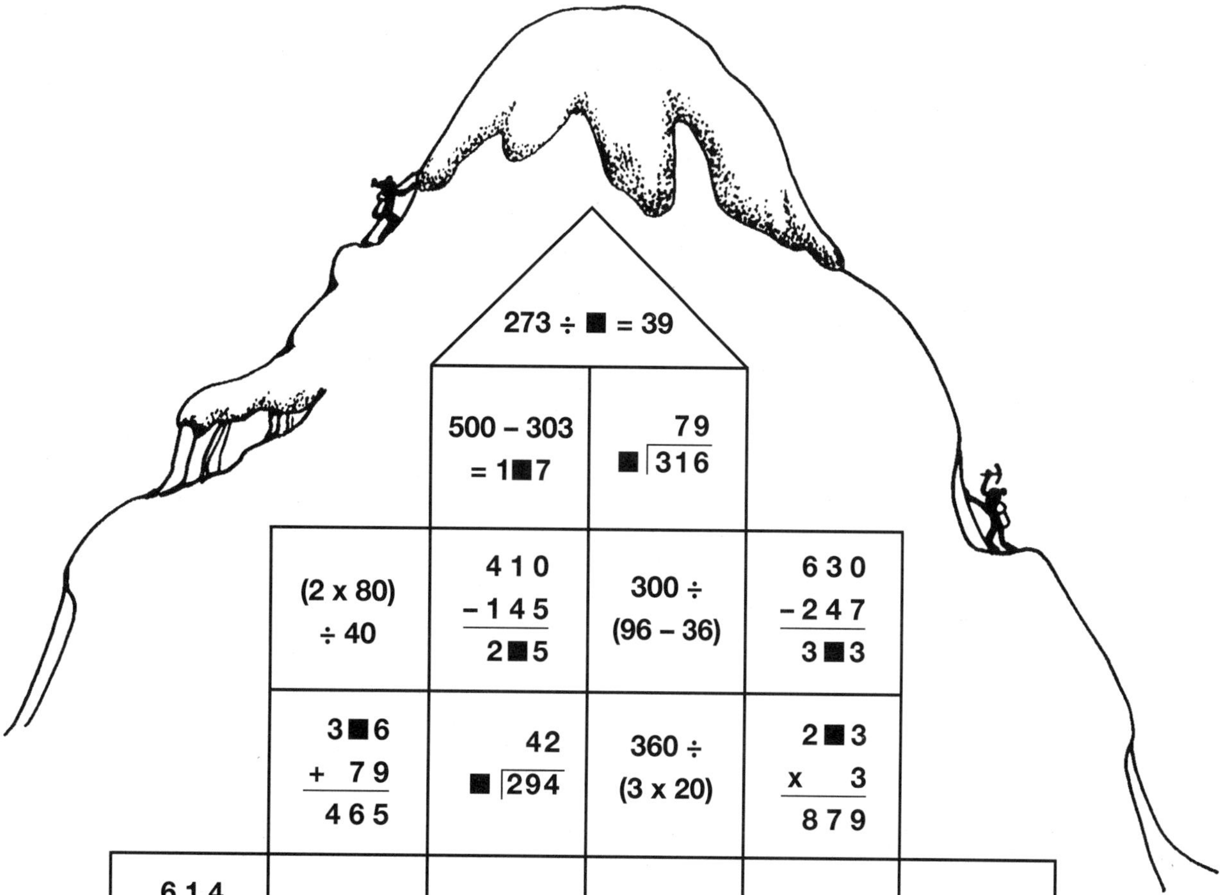

273 ÷ ■ = 39

500 – 303 = 1■7	79 ■)316

(2 x 80) ÷ 40	410 – 145 2■5	300 ÷ (96 – 36)	630 – 247 3■3
3■6 + 79 465	42 ■)294	360 ÷ (3 x 20)	2■3 x 3 879

614 – ■7 547	4 x ■3 = 292	(3 x 60) ÷ 20	145 ÷ ■ = 29	52 ■)208	63 ■)504
■5 x 4 = 340	45 ■)270	500 – 356 1■4	261 ÷ ■ = 29	210 ÷ (112 – 82)	3■7 x 3 1071

348 ÷ ■ = 58	3 x ■4 = 282	510 – 236 2■4	(4 x 40) ÷ 20	38 ■)190	104 x 6 62■	5■ x 6 = 336	(30 x 10) ÷ 60
601 – 146 4■5	200 ÷ (85 – 35)	7 x 3■ = 245	1■5 x 5 825	280 ÷ (8 x 5)	392 ÷ ■ = 49	33 ■)297	■6 x 3 138

Reach the Peak

Expert Level

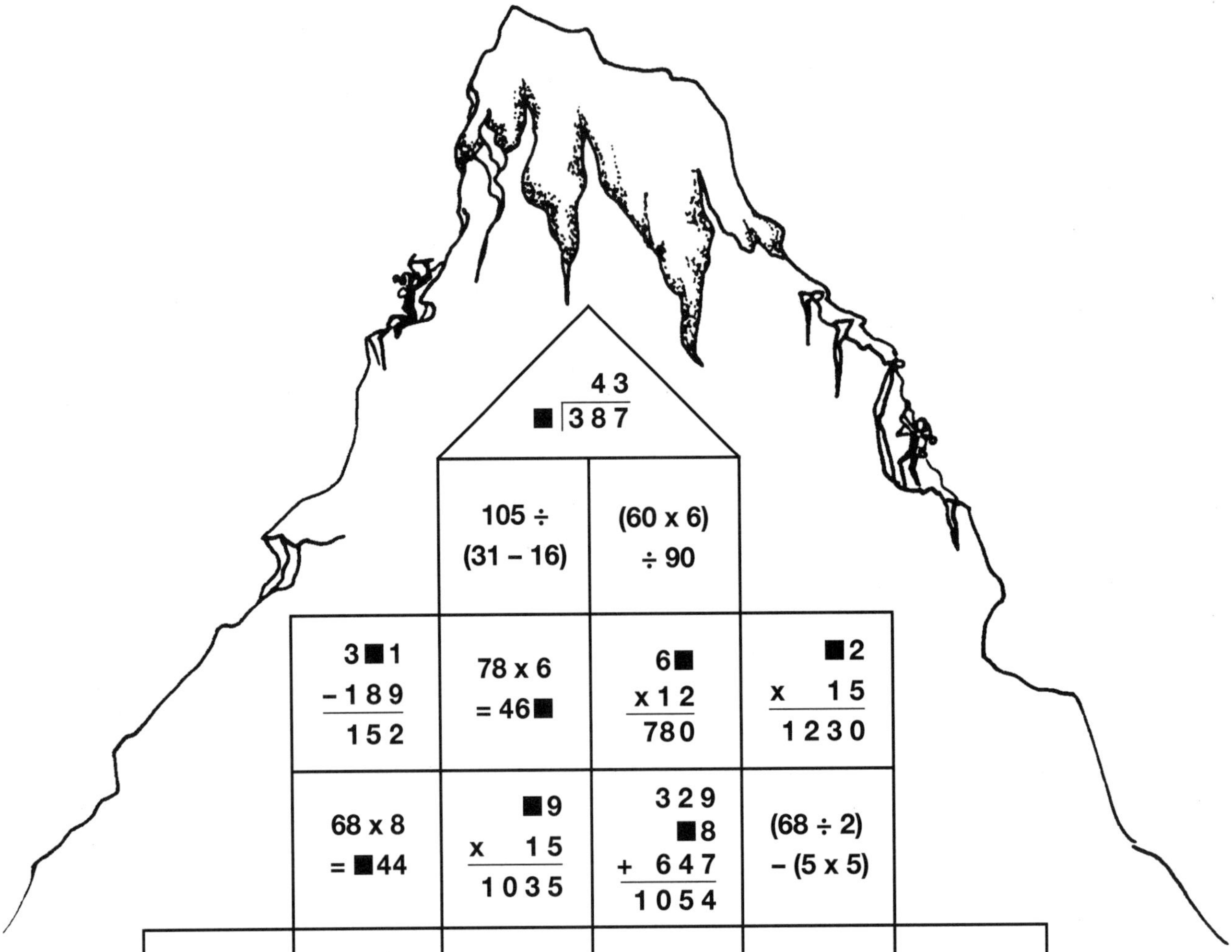

43 ■)387

105 ÷ (31 – 16)	(60 x 6) ÷ 90

3■1 – 189 152	78 x 6 = 46■	6■ x 12 780	■2 x 15 1230
68 x 8 = ■44	■9 x 15 1035	329 ■8 + 647 1054	(68 ÷ 2) – (5 x 5)

66 x 7 = ■62	58 x ■ 406	135 ÷ (32 – 17)	1012 – 8■5 147	36 ■)252	(15 x 30) ÷ 90
12,005 – ■543 3462	(164 ÷ 4) – (6 x 6)	54 ■)324	29 x 12 3■8	34 x ■ = 306	96 4)3■4

(136 ÷ 4) – (4 x 7)	27 x 11 2■7	121 x 6 = ■26	(4 x 60) ÷ 30	220 ÷ (50 – 6)	56 ■)224	1000 – 6■7 343	(3 x 140) ÷ 70
3652 + 7638 11,2■0	140 ÷ (81 – 46)	67 ■)335	86 x 7 = ■02	4010 – 298 3■12	128 ÷ (4 x 4)	18 x 20 ÷ 40	(8 x 4) – (140 ÷ 5)

Qualifying Equations

Topic: Mental Computation

Object: Create qualifying equations.

Groups: 2 players or pair players

Materials for each group

- *Qualifying Equations* recording sheet (for each player), p. 48
- 4 number cubes (1–6) (8 number cubes are ideal)
- scratch paper and pencil
- calculators (optional)

Tip *After students gain confidence playing the game, use special number cubes with higher numbers or add a fifth number cube.*

Directions

1. Each player rolls four number cubes and records the four numbers on his or her own recording sheet.
2. Each player works independently to use all four numbers to create an expression that results in a number that matches a description on his or her recording sheet.
3. All four rolled numbers must be used in the expression on the left side of the equation and must include one or two 2- or 3-digit numbers.
4. When a player's created problem matches a description on the recording sheet, the player records the complete equation on the line above that description.
5. When both players are ready, the completed equations are shared and verified as correct by the opposing player. If none of the remaining descriptions can be matched, nothing is recorded.
6. These same steps are followed for each round. Play ends when one player has created six complete equations to match each of the six descriptions.

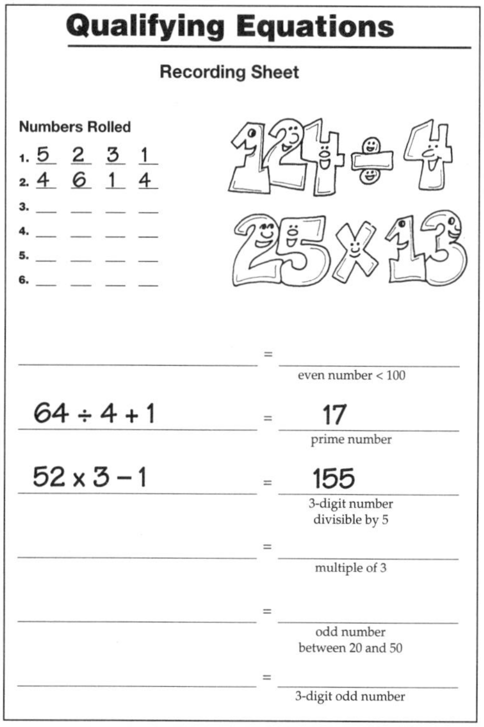

Qualifying Equations

Recording Sheet

Numbers Rolled

1. 5 2 3 1
2. 4 6 1 4
3. ___ ___ ___ ___
4. ___ ___ ___ ___
5. ___ ___ ___ ___
6. ___ ___ ___ ___

______ = ______ even number < 100

64 ÷ 4 + 1 = 17 prime number

52 x 3 − 1 = 155 3-digit number divisible by 5

______ = ______ multiple of 3

______ = ______ odd number between 20 and 50

______ = ______ 3-digit odd number

Making Connections

Promote reflection and make mathematical connections by asking:

- Which descriptions were harder to match? Why?
- What strategy helped you record on each turn?

Qualifying Equations

Recording Sheet

Numbers Rolled

1. ___ ___ ___ ___
2. ___ ___ ___ ___
3. ___ ___ ___ ___
4. ___ ___ ___ ___
5. ___ ___ ___ ___
6. ___ ___ ___ ___

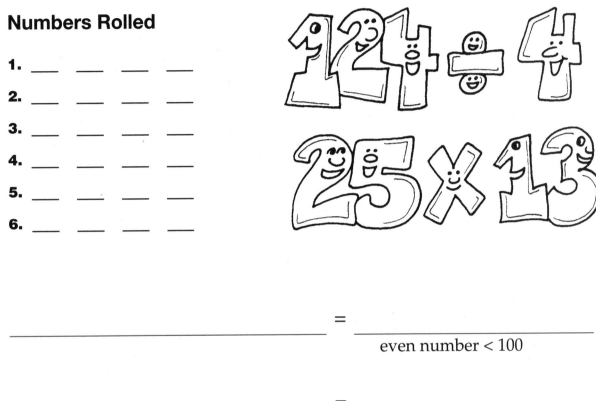

________________________ = ____________
even number < 100

________________________ = ____________
prime number

________________________ = ____________
3-digit number
divisible by 5

________________________ = ____________
multiple of 3

________________________ = ____________
odd number
between 20 and 50

________________________ = ____________
3-digit odd number

Date ____________ Name ____________________

Rearrange and Find I

You may cut out and use the number squares at the bottom of the page to help you solve the following problems.

1 2 4 5 Use 3 of these digits and any operation sign to make each equation true.

1. ___ ___ ◯ ___ = 36

2. ___ ___ ◯ ___ = 90

3. ___ ___ ◯ ___ = 27

4. ___ ___ ◯ ___ = 82

5. ___ ___ ◯ ___ = 208

6. ___ ___ ◯ ___ = 13

1 2 4 5 Use all 4 digits and any operation sign to make each equation true.

7. ___ ___ ◯ ___ ___ = 16

8. ___ ___ ◯ (___ ⊕ ___) = 125

9. ___ ___ ◯ (___ ⊖ ___) = 13

10. (___ ___ ◯ ___) ⊖ ___ = 209

11. ___ ___ ___ ◯ ___ = 608

12. ___ ___ ___ ◯ ___ = 1070

2 3 4 6 Use 3 of these digits and any operation sign to make each equation true.

13. ___ ___ ◯ ___ = 58

14. ___ ___ ◯ ___ = 78

15. ___ ___ ◯ ___ = 14

16. ___ ___ ◯ ___ = 40

17. ___ ___ ◯ ___ = 192

18. ___ ___ ◯ ___ = 258

2 3 4 6 Use all 4 digits and any operation sign to make each equation true.

19. ___ ___ ◯ ___ ___ = 28

20. (___ ___ ◯ ___) ⊕ ___ = 107

21. ___ ___ ◯ (___ ⊕ ___) = 182

22. (___ ___ ⊖ ___) ◯ ___ = 234

23. ___ ___ ___ ◯ ___ = 1448

24. ___ ___ ___ ◯ ___ = 2592

Create additional Rearrange and Find puzzles for classmates to try.

1	2	3	4	5	6

Date ____________ Name ____________

Rearrange and Find II

You may cut out and use the number squares at the bottom of the page to help you solve the following problems.

3 5 6 9 Use 3 of these digits and any operation sign to make each equation true.

1. ___ ___ ◯ ___ = 74

2. ___ ___ ◯ ___ = 195

3. ___ ___ ◯ ___ = 32

4. ___ ___ ◯ ___ = 87

5. ___ ___ ◯ ___ = 23

6. ___ ___ ◯ ___ = 567

---

3 5 6 9 Use all 4 digits and any operation sign to make each equation true.

7. ___ ___ ◯ ___ ___ = 37

8. (___ ___ ◯ ___) ⊗ ___ = 264

9. (___ ___ ⊖ ___) ◯ ___ = 435

10. (___ ___ ⊗ ___) ◯ ___ = 309

11. ___ ___ ◯ ___ ___ = 131

12. ___ ___ ⊗ (___ ◯ ___) = 105

13. ___ ___ ___ ◯ ___ = 5610

14. ___ ___ ___ ◯ ___ = 2370

---

2 4 7 8 Use 3 of these digits and any operation sign to make each equation true.

15. ___ ___ ◯ ___ = 108

16. ___ ___ ◯ ___ = 91

17. ___ ___ ◯ ___ = 592

18. ___ ___ ◯ ___ = 77

19. ___ ___ ◯ ___ = 39

20. ___ ___ ◯ ___ = 288

---

2 4 7 8 Use all 4 digits and any operation sign to make each equation true.

21. ___ ___ ◯ ___ ___ = 75

22. ___ ___ ◯ ___ ◯ ___ = 103

23. (___ ___ ◯ ___) ⨸ ___ = 33

24. (___ ___ ⊕ ___) ◯ ___ = 602

25. ___ ___ ⊗ (___ ◯ ___) = 54

26. ___ ___ ◯ (___ ⊖ ___) = 84

27. ___ ___ ___ ◯ ___ = 817

28. ___ ___ ___ ◯ ___ = 1148

Create additional Rearrange and Find puzzles for classmates to try.

1	2	3	4	5	6	7	8	9

Date ____________ Name ____________________

Related Cross-Number Puzzle

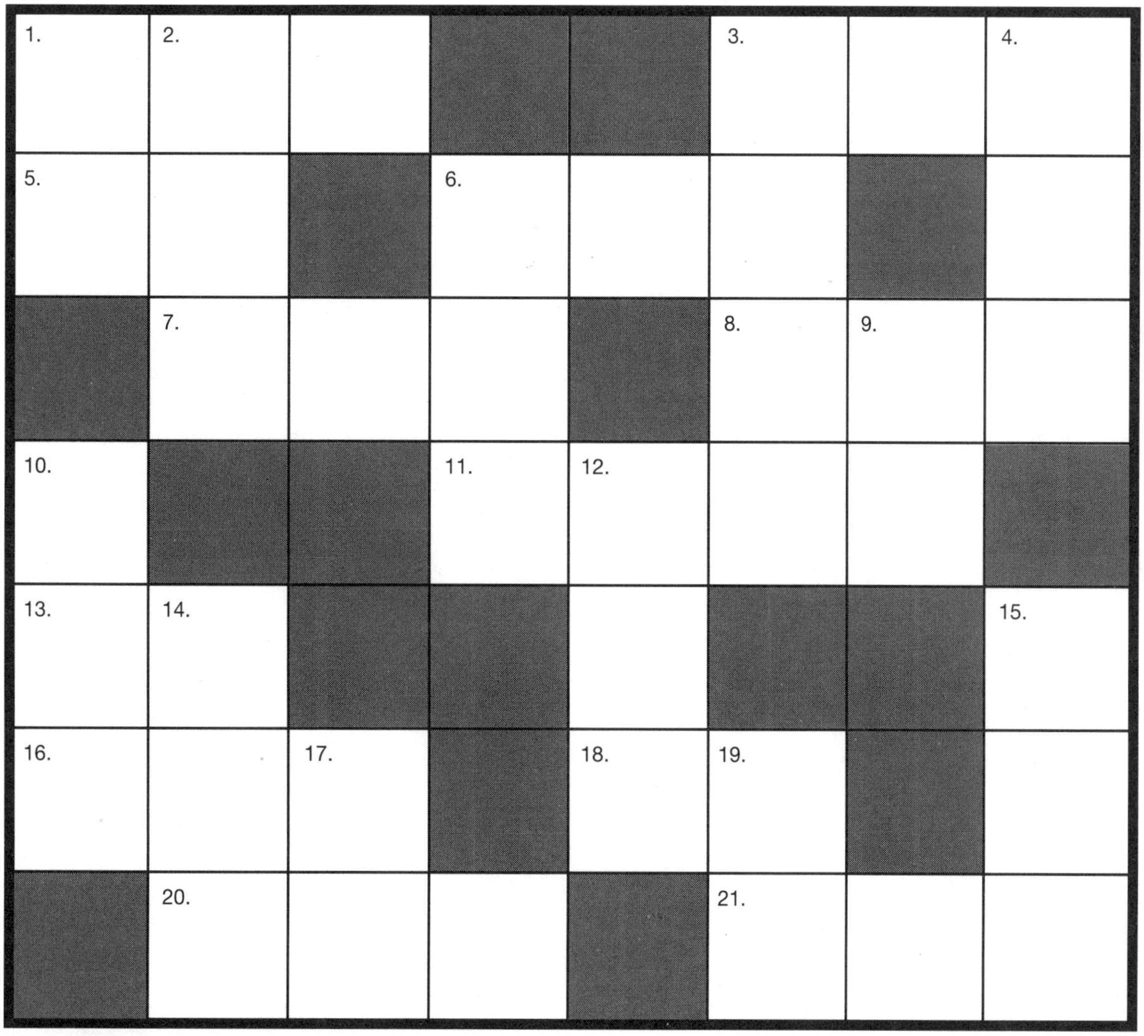

Across

1. 99 × 2
3. 1-Across + 207
5. 3-Across ÷ 5
6 5-Across ÷ 9 × 40
7. (6-Across – 250) × 2
8. 7-Across × 4 – 5
11. 2-Down × 5 + 322
13. 175 ÷ 5
16. 2-Down – 47
18. 9-Down × 10 ÷ 30
20. 1-Down × 50 + 58
21. 6-Across – 231

Down

1. 6-Across ÷ 20
2. 5-Across × 9 + 183
3. 12-Down × 5 – 22
4. 109 × 5
6. 2-Down ÷ 3
9. 6-Down – 232
10. 6-Down – 66
12. 4-Down + 277
14. 6-Across + 209
15. 4-Down – 176
17. 7-Across ÷ 4
19. (9-Down + 10) ÷ 2

Date ____________ Name ____________

Operation Fill Practice

Carefully insert operation signs in the following number sequences to produce accurate equations. The hints should help you find solutions. The first problem has been done for you.

Hint: Insert two operation signs, including at least one addition or subtraction sign.

1. (5 7 + 1 2 8) ÷ 5 = 37

2. 7 4 6 9 2 5 = 118

3. 9 3 8 9 8 4 = 210

4. 8 1 7 5 3 2 = 192

Hint: Insert two operation signs.

5. 6 3 5 1 4 6 = 64

6. 3 2 4 3 4 2 = 150

7. 1 9 2 8 1 2 = 288

Hint: Insert three operation signs in the following sequences.

8. 6 4 7 8 4 6 = 244

9. 9 2 1 0 3 2 = 315

10. 3 9 8 5 7 6 9 = 324

11. 1 0 2 6 4 3 5 = 12

Multiplication and Division

Assumptions Multiplication and division computation have previously been taught and reviewed, emphasizing understanding and building operation sense. Meaningful problems requiring multiplication and division have been posed and solved, allowing a variety of computational approaches. Mental estimation and computation are practiced regularly and frequently.

Section Overview and Suggestions

Sponge

Remainders Count p. 54

Finding Products p. 55

Product/Quotient Targets p. 56

These whole-class or small-group warm-ups are repeatable. They reinforce multiplication and division computation, while promoting mental estimation. Repeated use of all three sponges will ensure greater success with the Games and Independent Activities in this section.

Skill Checks

Products and Quotients Galore 1–6 pp. 57–59

The Skill Checks provide a way for parents, students, and you to see students' improvement with multiplication and division. Copies may be cut in half so that each check may be used at a different time. Be sure to have all students respond to the STOP, number sense task, before solving the ten problems.

Games

Unaligned Products pp. 60–61

Divisible Products pp. 62–63

3-2-1-Zero pp. 64–65

These repeatable, open-ended Games actively involve students in multiplying and dividing two- and three-digit numbers, while furthering operation sense. The *Finding Products* sponge is a good lead-up to *Unaligned Products,* while *Product/Quotient Targets* prepares students for *3-2-1-Zero.*

Independent Activities

Detect the Digits p. 66

Products Galore pp. 67–68

Finding Products Practice p. 69

Finding Quotients Practice p. 70

These activities motivate students to independently practice multiplication and division. Use of the *Finding Products* sponge will ensure greater success with *Finding Products Practice* and *Finding Quotients Practice,* which allow multiple solutions. Any of these activities can be replicated with new numbers to provide additional computation practice.

Remainders Count

Topic: Two and three-digit Quotients with Remainders

Object: Create division problems to generate high remainders.

Groups: Whole class or small group

Tip Operation sense and divisibility rules are further enhanced by having students seek the smallest possible remainders.

Materials

- transparent set of Digit Squares, p. 151
- set of Digit Squares for each participant
- scratch paper

Directions

1. The leader mixes the set of Digit Squares and displays this division format which is copied by the students.

   _) _ _ _

2. After the leader draws and announces four digits, each student locates the same four Digit Squares and arranges them to form a division problem.
3. The students record one or more division problems with complete answers, including the remainders.
4. The range of possible remainders is noted as students share their solutions with the class.
5. The students identify the arrangement(s) that generates the largest remainder.
6. The leader remixes all the Digit Squares, draws and announces four new digits, and asks the students to create a problem that generates the largest remainder.

   1 _) _ _ _

7. This process is continued with new digits and arrangements.
8. When students seem ready, the other division formats are displayed and used. The second format requires five Digit Squares.

   1 _) _ _ _ _

Making Connections

Promote reflection and make mathematical connections by asking:

- What digits are desirable for divisors? Please explain.
- What strategies helped you produce high remainders?

Finding Products

Topic: Mental Multiplication and Division

Object: Identify two factors whose product matches a given description.

Groups: Whole class or small group

Materials

- chalkboard or overhead projector

Directions

1. The leader displays nine two-digit numbers.
2. The leader describes a possible product.

   Example: "Find an even product under 1000."

3. Students identify two factors that produce a product that meets the description. Students are required to justify their selection of factors.

   Possible response to example: "18 and 49. 18 times 48 is less than 20 times 50 which equals 1000. 8 times 9 produces an even number."

4. The leader asks for other possible solutions to the same description.
5. The leader continues by announcing additional product descriptions.

   Possible descriptions for the same displayed choices:

   a 4-digit product
   a product under 500
   a product greater than 3000
   a product less than 1000 that's a multiple of 5
   an odd product between 1000 and 1200
   a product between 1200 and 2000 that's a multiple of 3
   a product under 1800 that ends in 6
   the smallest product that ends in 8
   the largest possible product that ends in 8
   a product between 1400 and 1600
   the largest odd-numbered product

Tips *Have students partner to create additional* Finding Products *factor choices and descriptions. For more practice, have students complete the* Finding Products Practice *sheet, p. 69.*

26	49	12
35	60	23
97	18	54

Making Connections

Promote reflection and make mathematical connections by asking:

- What approach helped you quickly locate possible factors?
- Which products are harder to mentally multiply? Please explain.

Product/Quotient Targets

Topic: Multiplication and Division

Object: Create problems to match given descriptions.

Groups: Whole class or small group

Materials

- transparent set of Digit Squares, p. 151
- container for Digit Squares
- scratch paper

Tip *Introduce the warm-up by announcing all five digits and having students arrange the digits to match the description.*

Directions

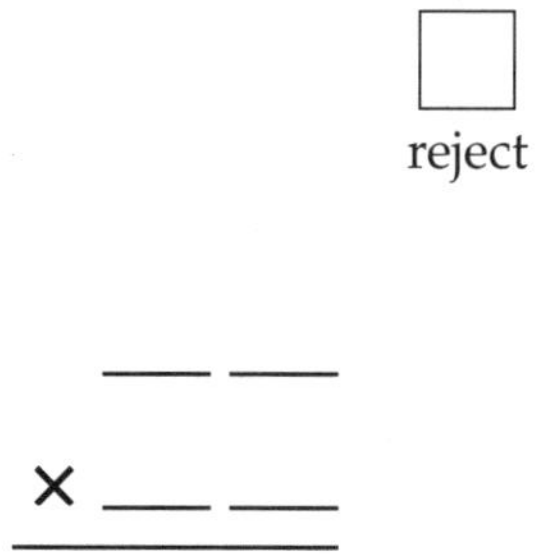

1. The leader displays a computation format which students copy. The leader then describes the characteristics of the desired product or quotient.

Examples: "Create a multiplication problem that produces a product greater than 2000," **or** "Create a division problem that produces a quotient less than 100."

2. The leader explains that five Digit Squares will be drawn and announced one at a time.

3. Each student records the digit as it is announced on one of the format lines. Since there are five digits and four lines, students discard one digit in the reject box. Students may not change or move the digits once they are recorded.

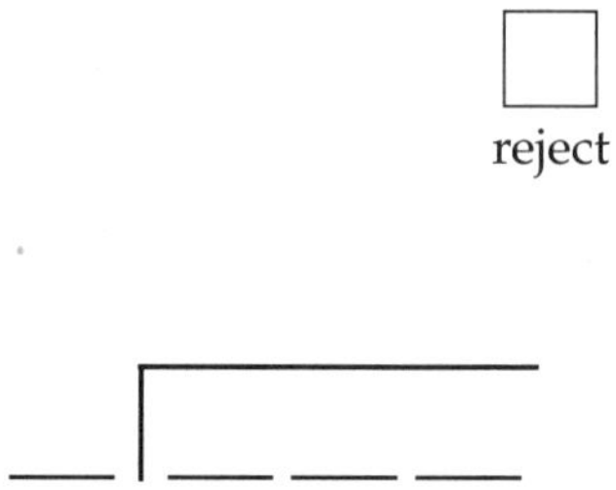

4. This process is followed until all five squares have been drawn. (Drawn squares are not returned to the container until the end of the round.)

5. After students have recorded the fifth digit, they solve their problems and determine which products (or quotients) match the given description.

6. Students continue with additional rounds of this activity.

Possible descriptions for future rounds:

an odd-numbered product

a quotient less than 300

a product between 1200 and 1700

an even-numbered quotient

7. If a more challenging format with more digits is used, draw enough Digit Squares to accommodate the format and the reject box.

Making Connections

Promote reflection and make mathematical connections by asking:

- How did you decide where to place certain digits?

Date ____________ Name ____________

Products and Quotients Galore 1

STOP Don't start yet. Star problems that may have answers less than 1000.

1. 648×7

2. 752×70

3. 36×24

4. 45×87

5. 412×35

6. $7\overline{)261}$

7. $60\overline{)434}$

8. $80\overline{)6086}$

9. $21\overline{)1197}$

10. $59\overline{)2834}$

Go On Use the digits 3, 4, 6, 7 and 8 to write a multiplication problem with the largest possible product. Describe your strategy.

Date ____________ Name ____________

Products and Quotients Galore 2

STOP Don't start yet. Star problems that may have answers greater than 20,000.

1. 895×8

2. 624×40

3. 52×34

4. 67×98

5. 526×43

6. $8\overline{)364}$

7. $70\overline{)575}$

8. $60\overline{)4924}$

9. $51\overline{)3264}$

10. $39\overline{)2068}$

Go On Use the digits 4, 5, 6, and 8 to write a division problem with the largest whole-number quotient. Describe your strategy.

Date ____________ Name ____________

Products and Quotients Galore 3

STOP Don't start yet. Star problems that may have odd answers.

1. 787×9

2. 537×60

3. 43×35

4. 38×79

5. 435×45

6. $6\overline{)315}$

7. $80\overline{)571}$

8. $70\overline{)3713}$

9. $31\overline{)2356}$

10. $49\overline{)3627}$

Go On Use +, –, × or ÷ to make each equation true.

(43 ◯ 27 ◯ 13) ◯ 14 = 82 (630 ◯ 18 ◯ 13) ◯ 12 = 4

Date ____________ Name ____________

Products and Quotients Galore 4

STOP Don't start yet. Star problems that may have four-digit answers.

1. 956×7

2. 874×80

3. 37×43

4. 76×89

5. 538×54

6. $9\overline{)578}$

7. $70\overline{)643}$

8. $90\overline{)5767}$

9. $61\overline{)3782}$

10. $39\overline{)2537}$

Go On Use the digits 2, 3, 5, 6, and 9 to write a multiplication problem with the smallest possible product. Describe your strategy.

Date ____________________ Name ________________________________

Products and Quotients Galore 5

STOP Don't start yet. Star each problem that may have answers with multiples of 10.

1. 794×8

2. 846×50

3. 54×25

4. 47×78

5. 324×35

6. $7\overline{)409}$

7. $80\overline{)652}$

8. $60\overline{)3426}$

9. $41\overline{)2993}$

10. $49\overline{)3530}$

Go On Use the digits 3, 5, 7, and 9 to write a division problem with the smallest whole-number quotient. Describe your strategy.

Date ____________________ Name ________________________________

Products and Quotients Galore 6

STOP Don't start yet. Star two problems that may have quotients with remainders.

1. 866×9

2. 659×90

3. 45×44

4. 89×97

5. 647×52

6. $8\overline{)533}$

7. $60\overline{)556}$

8. $80\overline{)5448}$

9. $71\overline{)4118}$

10. $59\overline{)3364}$

Go On Use +, –, × or ÷ to make each equation true.

(61 ○ 27) ○ 8 ○ 4 = 15 (35 ○ 17) ○ 16 ○ 26 = 32

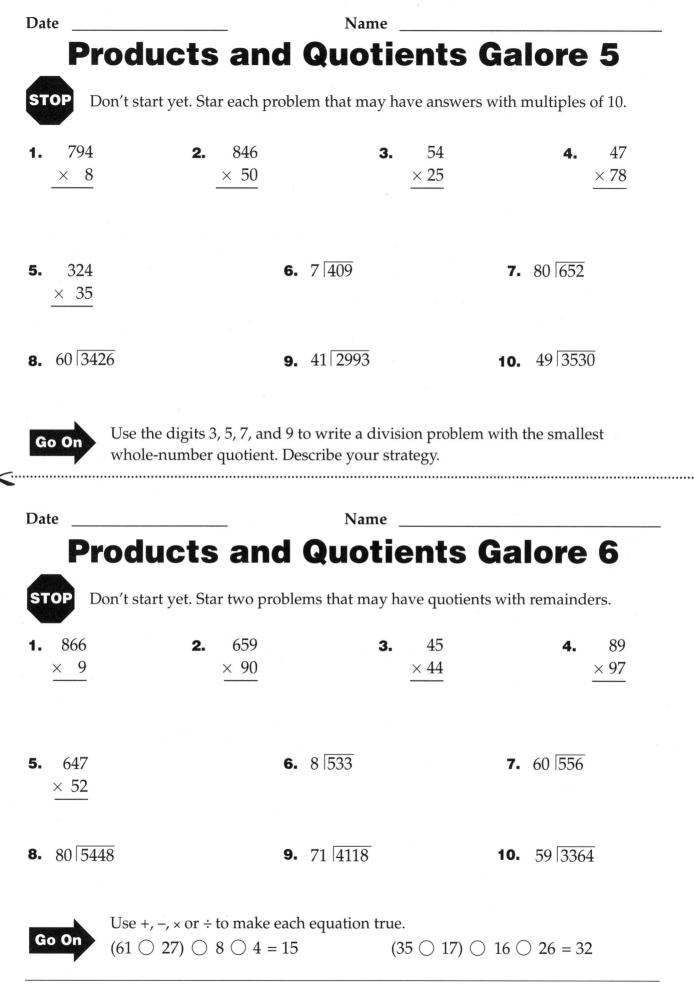

Unaligned Products

Topic: Mental Multiplication and Division

Object: Avoid covering three numbers in a row.

Groups: 2 players or pair players

Materials for each group

- *Unaligned Products A* gameboard, p. 61
- different kind of markers for each player
- calculator (optional)

Tip *Consider using the more challenging factor choices on* Unaligned Products B *gameboard, p. 61.*

Directions

1. The first player selects and announces two factors (one from each row in the "Choices" box). After the factor choices are announced, players compute the product with paper and pencil or with a calculator. The first player covers the resulting product on the gameboard with her or his marker.
2. The other player selects and announces two factors whose product is computed by both players. The resulting product is covered by that player's marker.
3. Players continue alternating turns, trying to avoid aligning three markers in a row. Players must select two factors that result in a product not yet covered. This might require more than one attempt.
4. The first player forced to have three markers in a row horizontally, vertically, or diagonally loses.

8847	2168	1652	4878
3932	944	3213	2710
1180	2124	1836	6881
4915	3794	4131	2295

Making Connections

Promote reflection and make mathematical connections by asking:

- What strategy helped you avoid lining up your markers in a row?

Unaligned Products A

8847	2168	1652	4878
3932	944	3213	2710
1180	2124	1836	6881
4915	3794	4131	2295

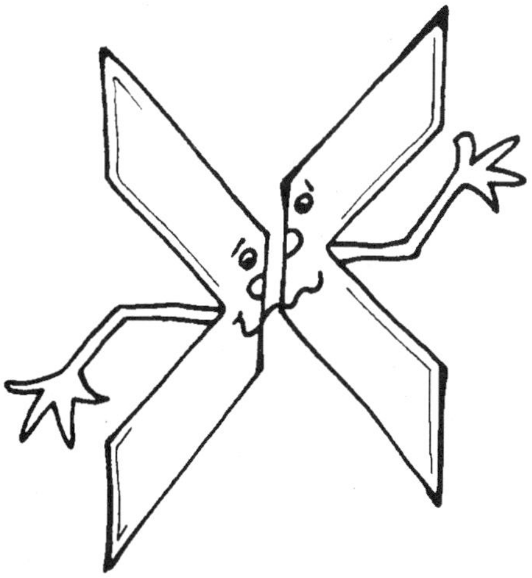

Choices

4	5	7	9
236	459	542	983

Unaligned Products B

6408	2148	6360	10,680
4272	7240	2896	5874
3982	4344	1969	15,900
9540	1432	3580	8745

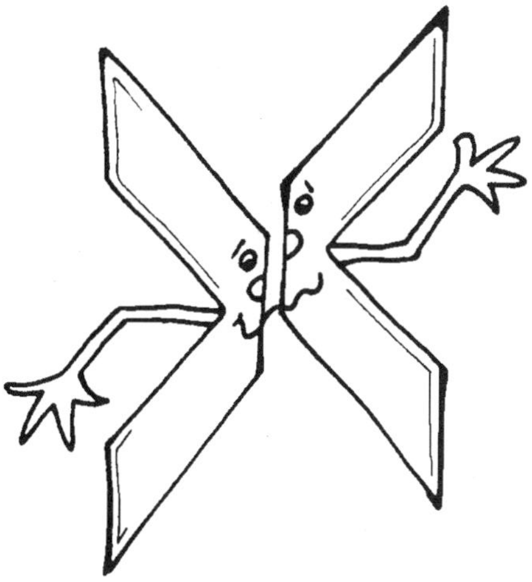

Choices

8	11	12	20
179	362	534	795

Divisible Products

Topic: Multiplication and Division

Object: Create expressions with large values.

Groups: 2 players or pair players

Materials for each group

- *Divisible Products* score sheet, p. 63
- 3 number cubes (1–6)
- special number cube (4–9), p. 154

Tip As students get more confident with this game, consider substituting more 4–9 number cubes for the 1–6 cubes.

Directions

1. The first player rolls the four cubes to generate four digits.
2. The player determines how best to use the four rolled numbers. Two of the generated numbers form a two-digit number which is multiplied by a third number and divided by the remaining number.
3. When a player decides how to use the four rolled numbers, the resulting equation is recorded on the score sheet. If 1 is a number choice, it is not allowed to be the divisor.

   Example: 6, 1, 6, and 5 are rolled. The player decides to make the following equation, $61 \times 6 \div 5 = 73\ r\ 1$.

   (6 1 × 6) ÷ 5 = 73 r1 (quotient) 70 (score*)

4. If the quotient is a whole number, that number is the score. If, as in the example above, there is a remainder in the quotient, the player's score is the whole number part of the quotient less three times the remainder. (In the above example, the player earns 70 points.) The score for each player's round is recorded.
5. The next player has two choices. The player may use the same four numbers generated by the previous player and create a different high scoring equation or the player may roll the four number cubes, determine an equation, and record that equation and score.
6. Players continue to alternate turns until each player has recorded five equations. The scores for each player's five equations are totaled to determine each player's final score. High score wins.

Making Connections

Promote reflection and make mathematical connections by asking:

- When and why did you decide to opt for equations with remainders?
- Where did you usually place lower numbers?
- If you chose to use the numbers previously generated, what was the reasoning behind your choice?

Divisible Products

(___ ___ × ___) ÷ ___ = ______ quotient ______ score*

(___ ___ × ___) ÷ ___ = ______ quotient ______ score*

(___ ___ × ___) ÷ ___ = ______ quotient ______ score*

(___ ___ × ___) ÷ ___ = ______ quotient ______ score*

(___ ___ × ___) ÷ ___ = ______ quotient ______ score*

______ TOTAL SCORE

*score = whole-number quotient minus remainder × 3

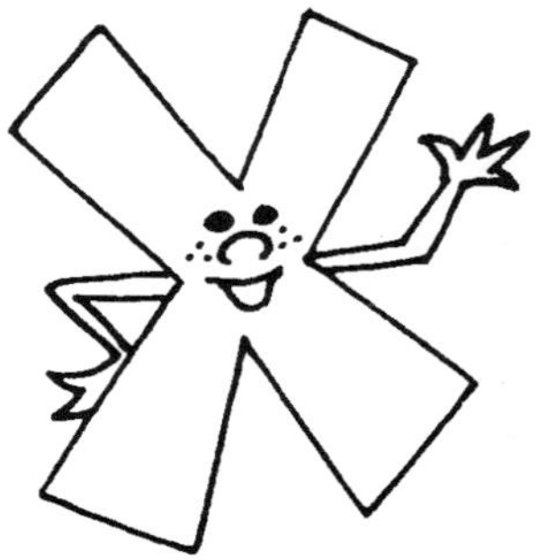

Divisible Products

(___ ___ × ___) ÷ ___ = ______ quotient ______ score*

(___ ___ × ___) ÷ ___ = ______ quotient ______ score*

(___ ___ × ___) ÷ ___ = ______ quotient ______ score*

(___ ___ × ___) ÷ ___ = ______ quotient ______ score*

(___ ___ × ___) ÷ ___ = ______ quotient ______ score*

______ TOTAL SCORE

*score = whole-number quotient minus remainder × 3

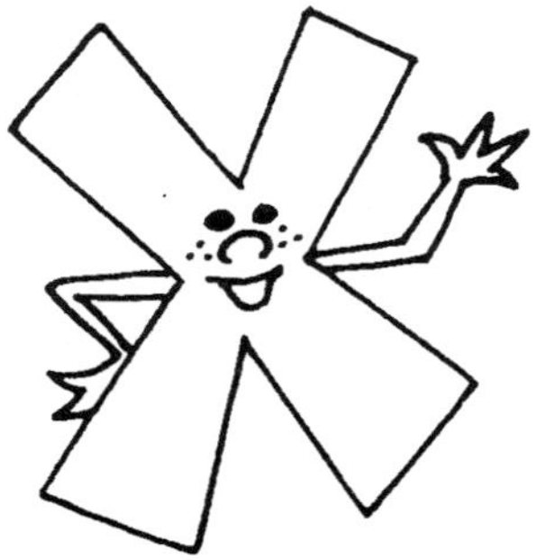

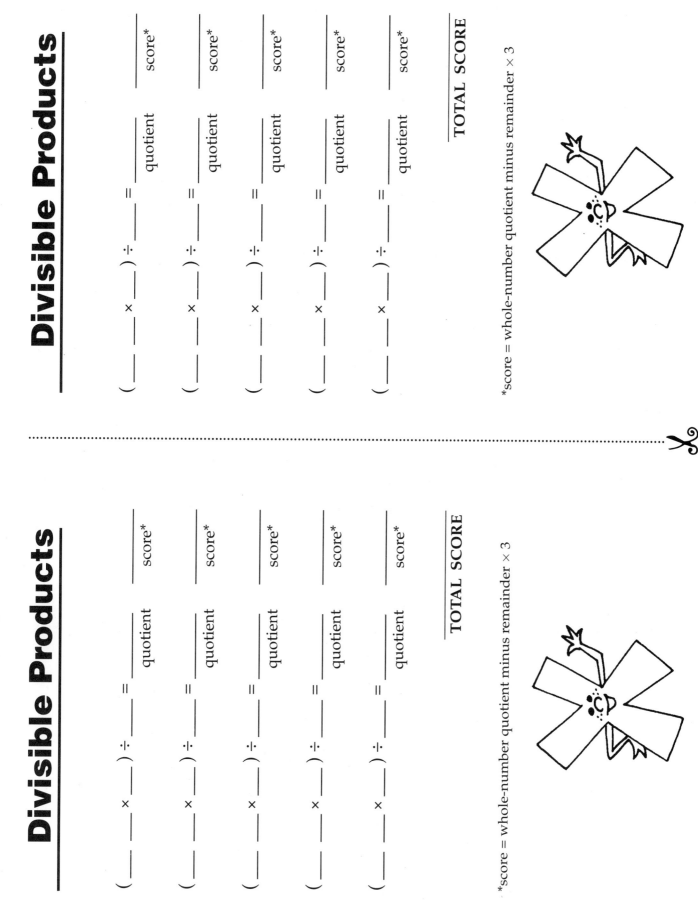

3-2-1-Zero

Topic: Multiplying and Dividing 2- and 3-digit numbers

Object: Create qualifying problems.

Groups: 2–4 players

Materials

- *3-2-1-Zero* recording sheet (for each player), p. 65
- 2 sets of Digit Cards, p. 150
- scratch paper and pencils

Tip *If students gain competence with this game, require players to draw and use five cards.*

Directions

1. One player distributes four Digit Cards to each player. Each player attempts to arrange all her or his cards to form a problem whose answer matches the description for the round.
2. If a player creates a problem that matches Round 1's description, she or he records the problem and answer. When enough time has passed, players share their problems or attempts. If a player's solution is accepted, the player records 3 points. If a player can use the same four digits to create a different yet qualifying problem, he or she is awarded one bonus point.

   Example: One player receives 2, 3, 3 and 5 and cannot form a problem. Another player receives 2, 3, 8 and 4 and forms two problems, 834 ÷ 2 and 832 ÷ 4, earning four points.

Description	Problem(s) & answer	Points
Round 1 whole number quotient with no remainder	834 ÷ 2 = 417 832 ÷ 4 = 208	4
Round 2 product divisible by 3	27 x 15 405	3

3. Any player who does not find a problem for Round 1's description makes a second attempt with the next four cards dealt. If a player succeeds in matching Round 1's description on this second attempt, she or he earns two points. (Success on a third attempt earns only one point.) If a player is still unable to find a qualifying problem after three attempts, the player records a zero and proceeds to Round 2.
4. The used Digit Cards are remixed and redistributed after every round.
5. Players continue to create multiplication or division problems with answers to match each round's description. Since players are required to go in order through the rounds, often players are creating problems to match different descriptions.
6. Scoring and playing ends when one player records points for Round #6. Players total their points and the player with the most points wins.

Making Connections

Promote reflection and make mathematical connections by asking:

- Which products and quotients were harder to find? Why?
- What approach helped you find possible solutions?

3-2-1-Zero

Description	Problem(s) & answer	Points
Round 1 whole number quotient with no remainder		____
Round 2 product divisible by 3		____
Round 3 quotient greater than 100 and divisible by 2 (remainders allowed)		____
Round 4 odd product		____
Round 5 quotient divisible by 5 (remainders allowed)		____
Round 6 product between 1000–1500		____

Total Points: ____

3-2-1-Zero

Description	Problem(s) & answer	Points
Round 1 whole number quotient with no remainder		____
Round 2 product divisible by 3		____
Round 3 quotient greater than 100 and divisible by 2 (remainders allowed)		____
Round 4 odd product		____
Round 5 quotient divisible by 5 (remainders allowed)		____
Round 6 product between 1000–1500		____

Total Points: ____

Date ______________ Name ______________________________

Detect the Digits

Fill in the missing digits to complete the problems. Use the clue to check your answers for each row of problems.

1. 384 × 9 = 34□6

2. 67□ × 7 = 4□25

3. 829 × □ = 4□74

4. 457 × □0 = 18,□8□

Clue: The sum of the missing digits equals two more than three dozen.

---

5. 7 ⟌ □3□ = 76

6. 4 ⟌ 18□6 = □59

7. 8 ⟌ 38□8 = □7□

8. 30 ⟌ 1□90 = 5□

Clue: The sum of the missing digits equals 2 × 2 × 2 × 2 × 2.

---

9. 26 × 1□ = □38

10. 89 × □0 = □670

11. □4 × 41 = 30□4

12. □33 × 21 = 11,□□3

Clue: The sum of the missing digits equals the number of hours in a day and a half.

---

13. 12 ⟌ 5□0 = 4□

14. 2□ ⟌ 556□ = 1□2

15. □3 ⟌ 182□ = □9

16. 51 ⟌ □□97 = 8□ *r*13

Clue: The sum of the missing digits equals the number of years in three score.

Challenge: Make up some problems with missing digits for your classmates to solve. Try to include a clue.

Date ____________ Name ____________________

Products Galore I

Arrange the listed factors to produce the indicated products. The first two are started for you.

1. Factors: 9, 10, 11, 12

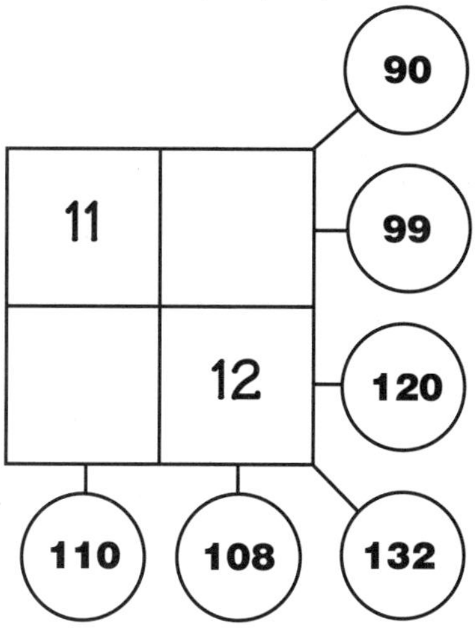

2. Factors: 11, 12, 13, 14

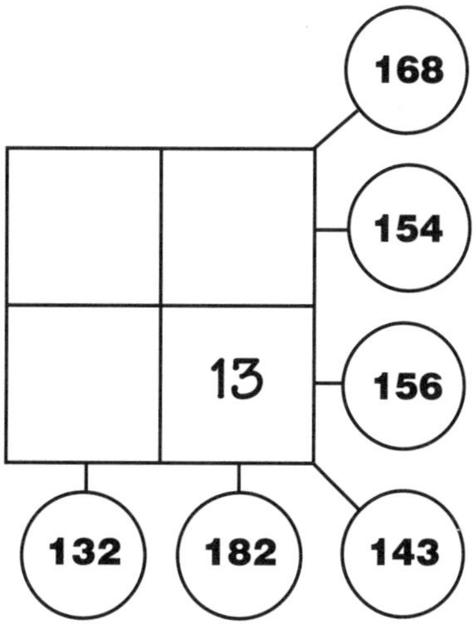

3. Factors: 12, 13, 14, 15

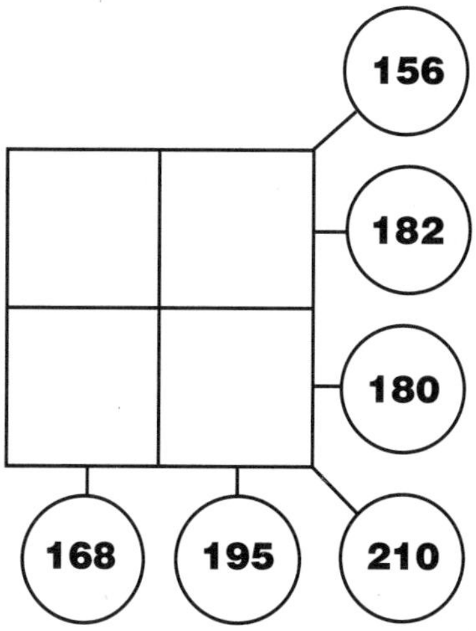

4. Factors: 15, 16, 17, 18

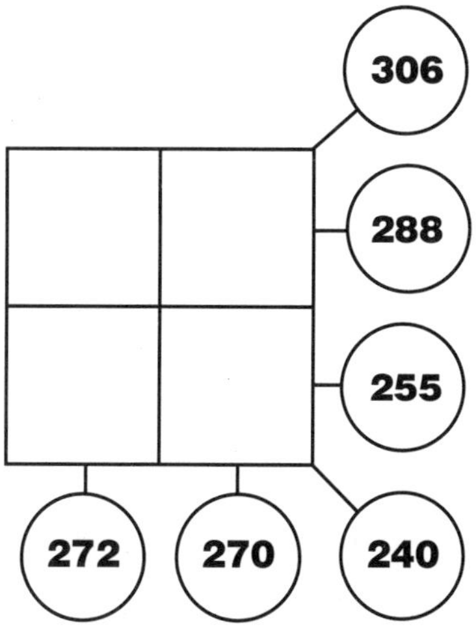

5. Factors: 7, 20, 24, 29

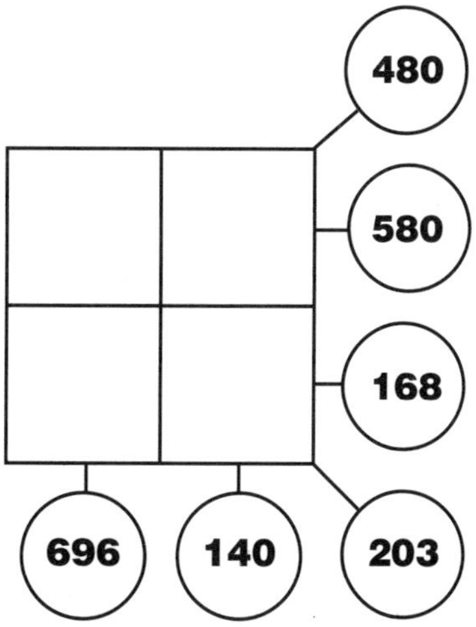

6. Factors: 8, 21, 23, 25

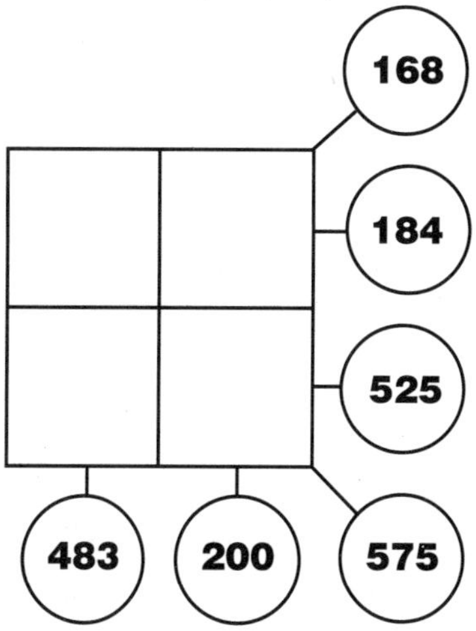

7. Factors: 9, 22, 26, 27

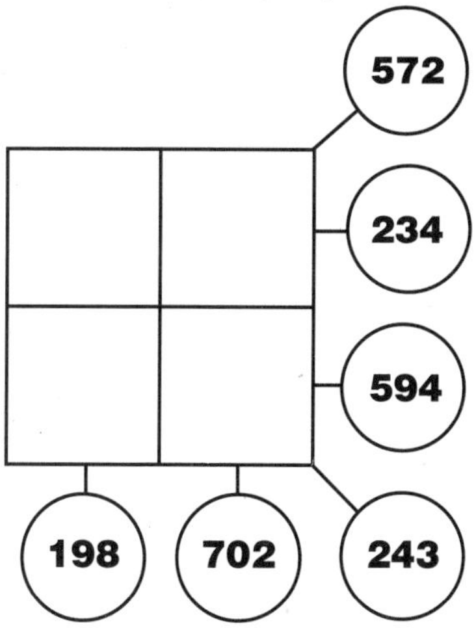

8. Factors: 21, 22, 23, 24

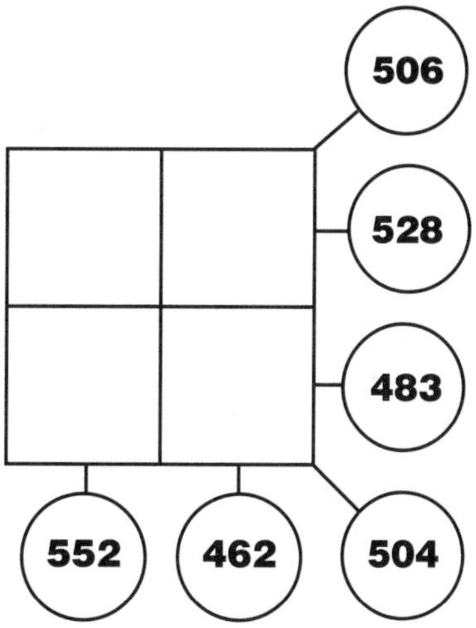

9. Factors: 6, ?, 22, 25

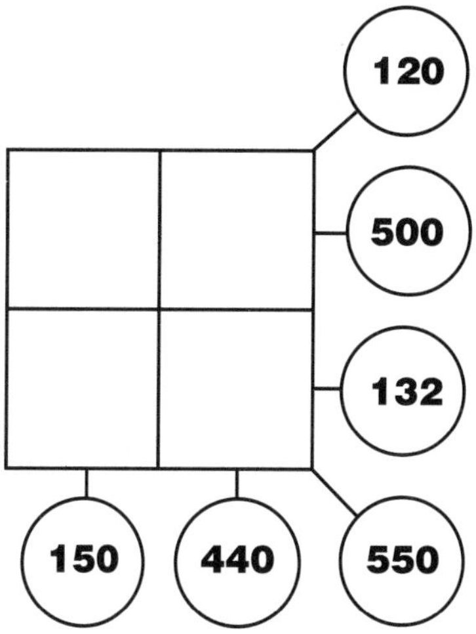

Date ____________ Name ____________________

Products Galore II

Arrange the listed factors to produce the indicated products.

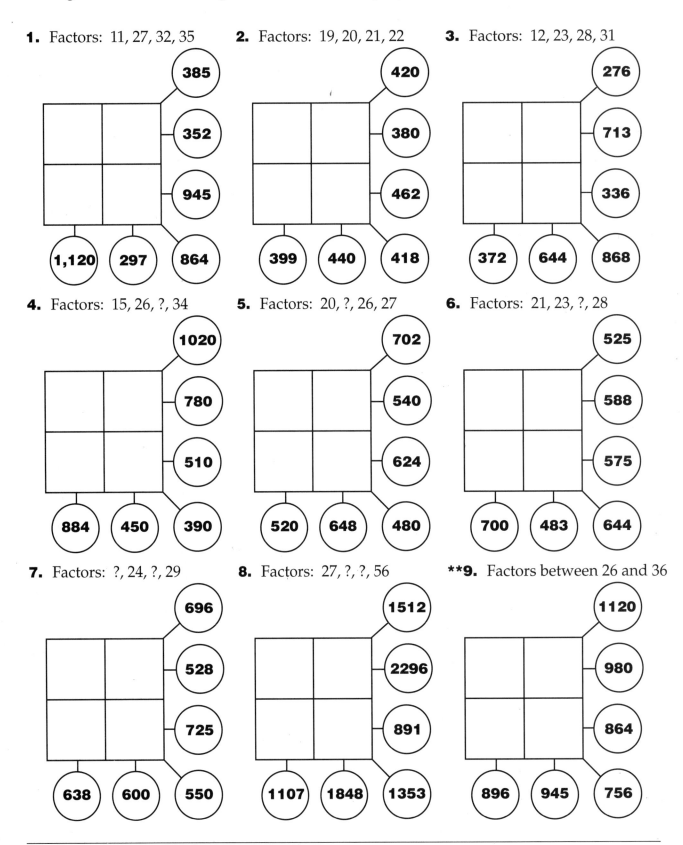

1. Factors: 11, 27, 32, 35

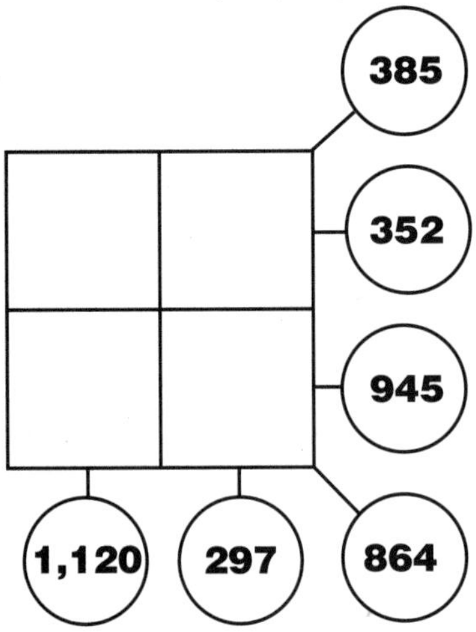

2. Factors: 19, 20, 21, 22

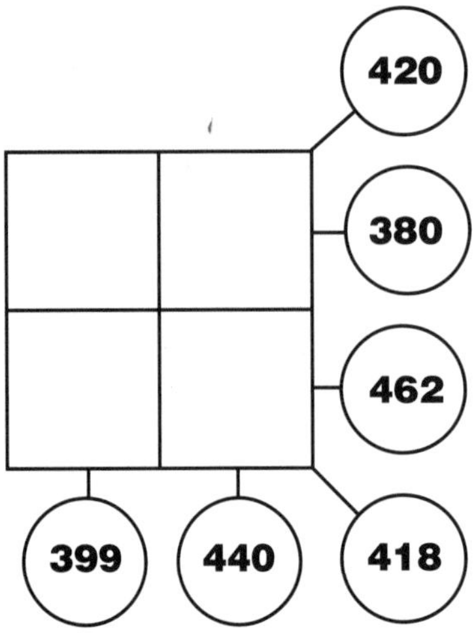

3. Factors: 12, 23, 28, 31

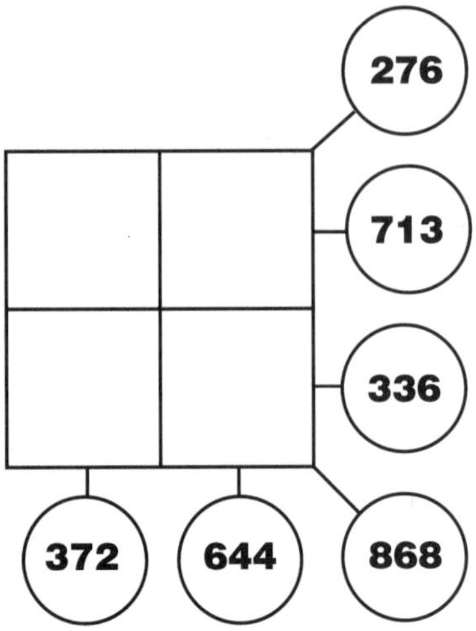

4. Factors: 15, 26, ?, 34

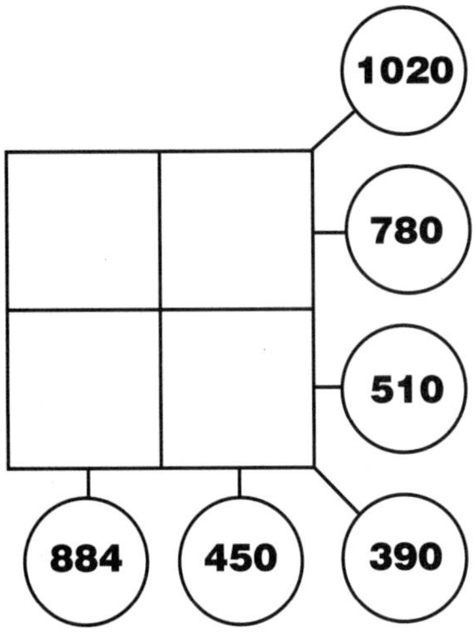

5. Factors: 20, ?, 26, 27

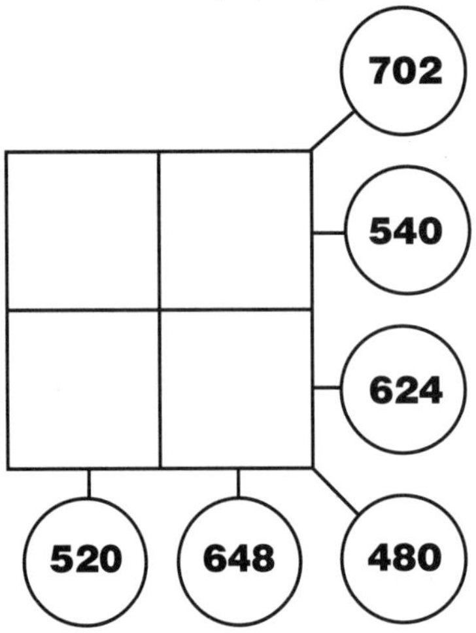

6. Factors: 21, 23, ?, 28

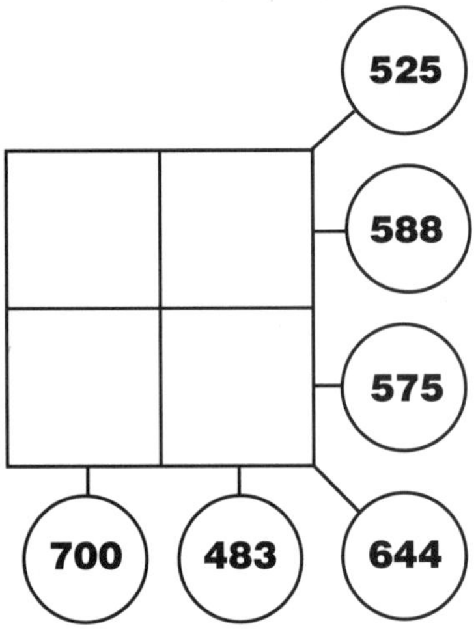

7. Factors: ?, 24, ?, 29

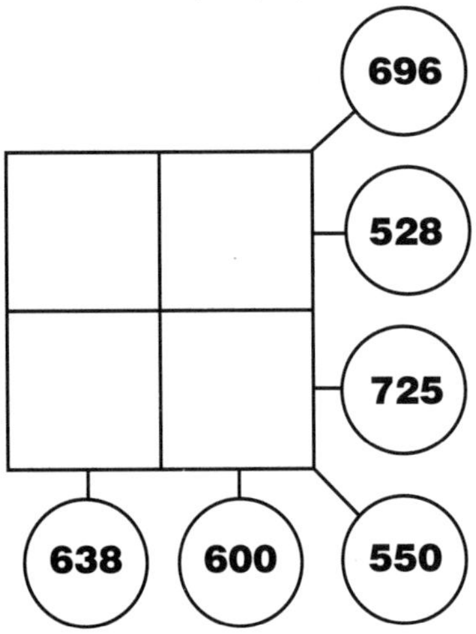

8. Factors: 27, ?, ?, 56

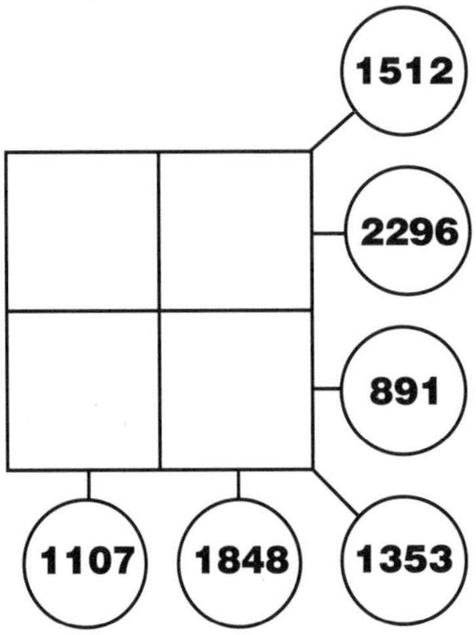

****9.** Factors between 26 and 36

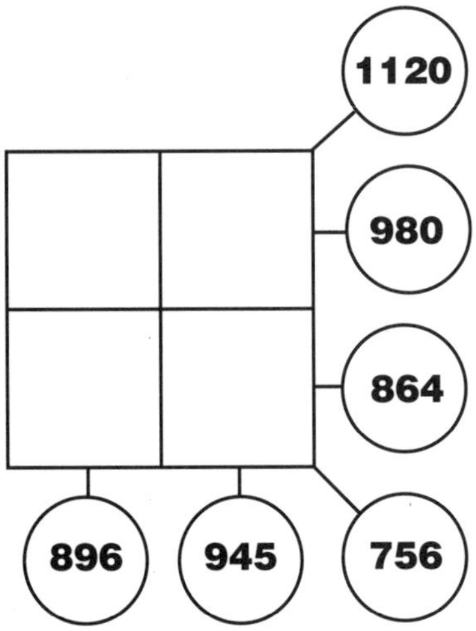

Date ____________ Name ____________________

Finding Products Practice

Match the descriptions below by using these factors to create multiplication problems and answers.

19	**21**	**38**	**42**	**70**	**85**

1. ______________________
even product between 700 and 800

2. ______________________
odd product

3. ______________________
4-digit product that's a multiple of 3

4. ______________________
product between 1000 and 2000 that's a multiple of 10

5. ______________________
smallest possible even product

6. ______________________
even product greater than 5000

Match the descriptions below by using these factors to create multiplication problems and answers.

17	**22**	**34**	**49**	**56**	**63**	**98**

7. ______________________
largest possible even product

8. ______________________
product between 1500 and 2000

9. ______________________
smallest possible odd product

10. ______________________
odd product greater than 3000

11. ______________________
even product between 2000 and 2500

12. ______________________
even product less than 500

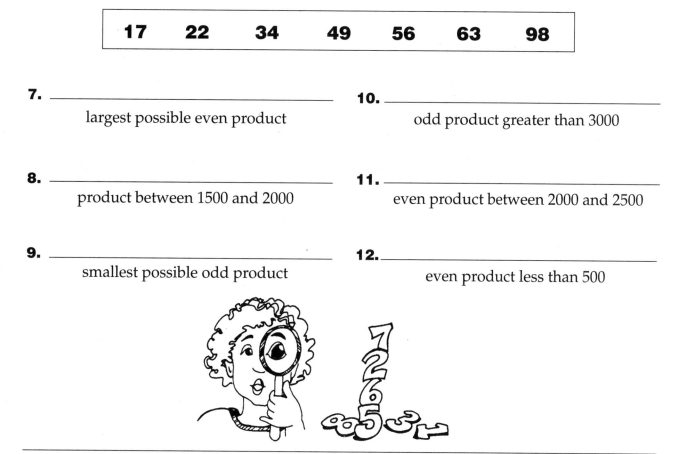

Date ____________________ Name ________________________________

Finding Quotients Practice

Match the descriptions below by using these divisors and dividends to create division problems and solutions.

Divisors			
15	23	36	60

Dividends				
1081	915	2052	180	2580

1. ______________________________
quotient less than 10 with no remainder

2. ______________________________
largest quotient

3. ______________________________
2-digit odd quotient with no remainder

4. ______________________________
quotient between 15 and 20

5. ______________________________
quotient with a remainder over 14

6. ______________________________
quotient with a remainder of 1

Match the descriptions below by using these divisors and dividends to create division problems and solutions.

Divisors			
18	39	45	54

Dividends				
594	1080	1248	486	2052

7. ______________________________
smallest quotient

8. ______________________________
odd quotient with no remainder

9. ______________________________
quotient with an odd remainder

10. ______________________________
quotient between 25 and 35 with no remainder

11. ______________________________
quotient with a remainder over 25

12. ______________________________
largest quotient with a remainder

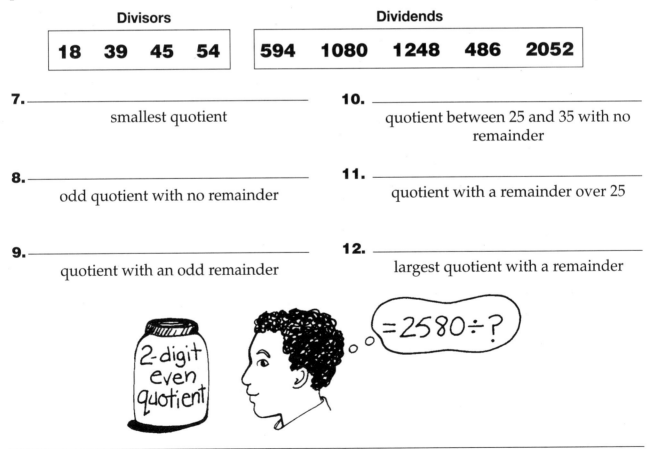

Fractions

Assumptions Fraction operations have previously been taught and reviewed, emphasizing understanding and building fraction sense. Concrete objects and visual models, such as fraction circles or rectangles and grids, have been used extensively.

Section Overview and Suggestions

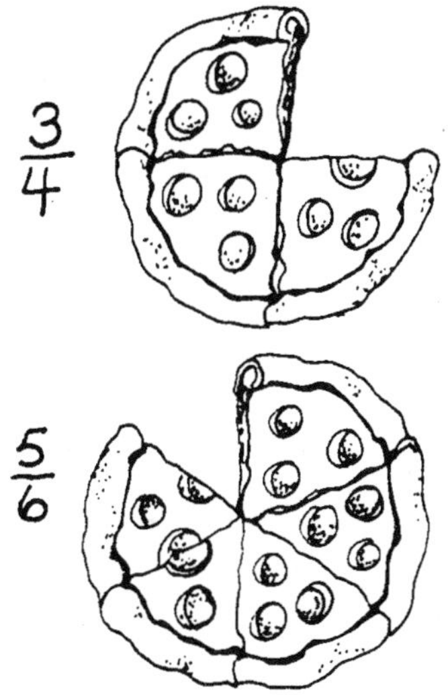

Sponges

Spinning Fractions pp. 72–73

Target Fractions pp. 74–75

Seeking Fractions pp. 76–77

These open-ended, repeatable, whole class or small group warm-ups require mental computation and reinforce all fraction operations. Repeated use of *Spinning Fractions* will ensure greater success with *Ordered Fractions.* In fact, use of all sponges will increase students' confidence with this section's Games and Independent Activities.

Skill Check

Partial Possibilities 1–6 pp. 78–80

The Skill Checks provide a way for parents, students, and you to see students' improvement with fraction computation. Copies can be cut in half so that each check may be used at a different time. Remember to have all students respond to the STOP, number sense task, before solving the ten problems.

Games

Ten Tallies Win pp. 81–82

Ordered Fractions pp. 83–84

Fraction Arrangements pp. 85–87

These open-ended and repeatable Games actively involve students in creating fractions and mentally comparing fractions. Use of the above Sponges will prepare students for *Fraction Arrangements,* which reinforces adding and subtracting fractions (Gameboard A), as well as multiplying and dividing fractions (Gameboard B).

Independent Activities

If ... Then ... p. 88

Digits to Fractions pp. 89–90

Fraction Choices pp. 91–92

If ... Then ... provides a good assessment of students' fraction sense. *Digits to Fractions* and *Fraction Choices* involve students in independent practice of fraction operations. Students mentally compute many problems as they seek expressions that work. Both of these Independent Activities provide two versions, allowing only practice of addition and subtraction on the easier version.

Spinning Fractions

Topic: Computing and Ordering Fractions

Object: Record fractions in an ascending sequence.

Groups: Whole class or small group

Materials

- transparency of *Spinning Fractions* activity form, p. 73
- transparent set of Digit Squares (0, 5, 7 & 9 removed), p. 151
- paper clip and pencil for spinner
- scratch paper for each participant

Tip *Once students gain familiarity with this warm-up, require them to select their fraction choice before the spin.*

Directions

1. On a blank sheet of paper, have each student draw six horizontal recording lines similar to those on the activity form. Inform students that eventually six fractions need to appear in order from least to greatest on these lines and that once they record a fraction, the fraction may not be moved.
2. The leader spins the spinner to determine how a fraction will be changed once it has been selected.
3. The leader draws and displays two Digit Squares. Students identify the two possible fraction choices using these digits. Students work independently (or in pairs, if preferred) to determine which digit to place in the numerator and which digit to place in the denominator.
4. Students work independently (or in pairs) to change their fraction choice according to the displayed spin.
5. After students agree on the computed fraction, they record it on one of the six lines. (The displayed Digit Squares are returned.)
6. The process of spinning, drawing two of the six Digit Squares, agreeing on a fraction, and cautiously recording the resulting fraction, continues until some students successfully order six of the resulting fractions. (If a resulting fraction will not fit a student's sequence, it is not recorded. Equivalent fractions are allowed to be recorded next to each other.)
7. Students with six ordered fractions share their results. (Students appreciate additional rounds once they realize the range of the possible resulting fractions.)

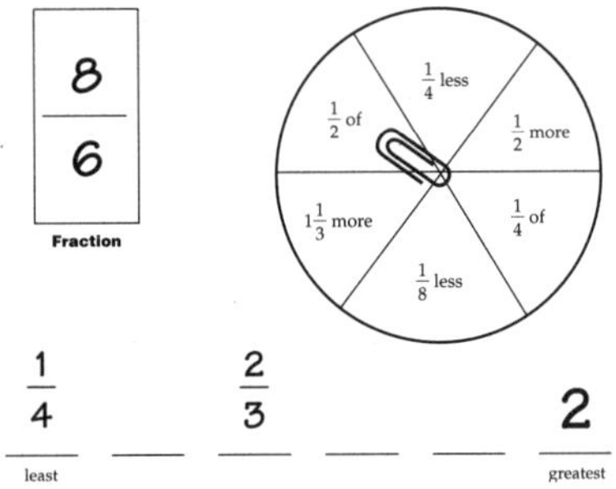

Making Connections

Promote reflection and make mathematical connections by asking:

- What helped you determine where to place your fractions?
- How could this activity be modified to allow greater success by all?

Spinning Fractions

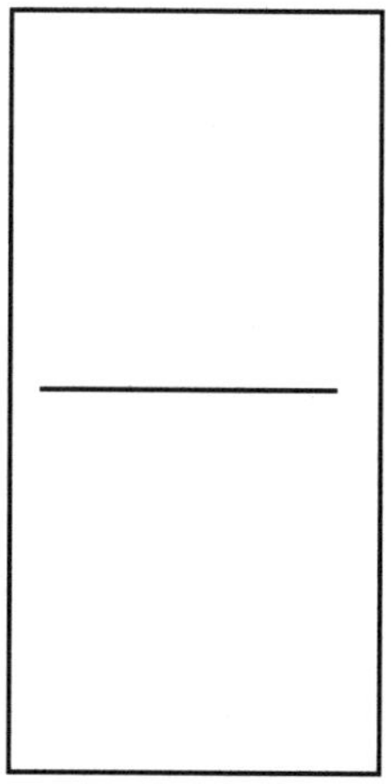

Fraction

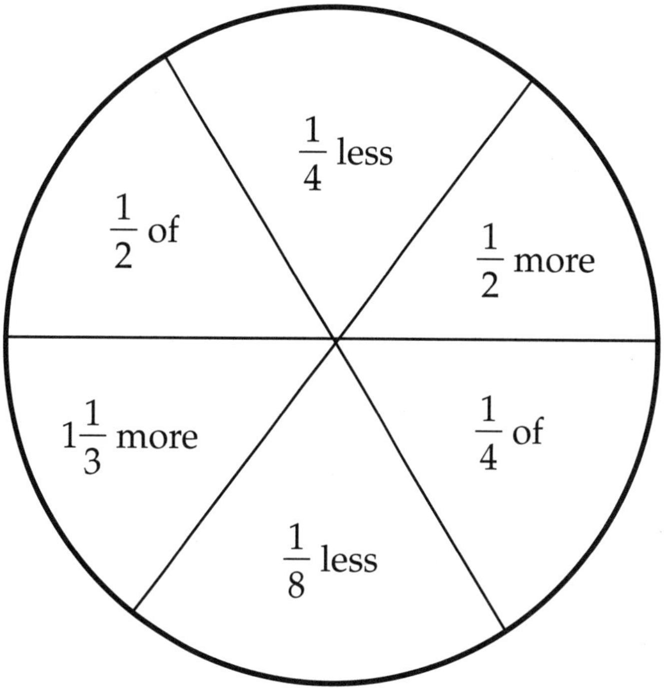

________ ________ ________ ________ ________ ________

least greatest

Target Fractions

Topic: Computing with Fractions

Object: Create fractions that generate a sum, difference, product, or quotient equal or close to a displayed target fraction.

Groups: Whole class or small group

Tip *Once students become familiar with this warm-up, have students total their differences for each round to generate comparative scores.*

Materials

- transparency of *Target Fractions* activity form, p. 75
- 2 sets of transparent Digit Squares (0's removed), p. 151
- 2 sets of Digit Squares for each participant
- paper clip and pencil
- blank sheet for each participant

Directions

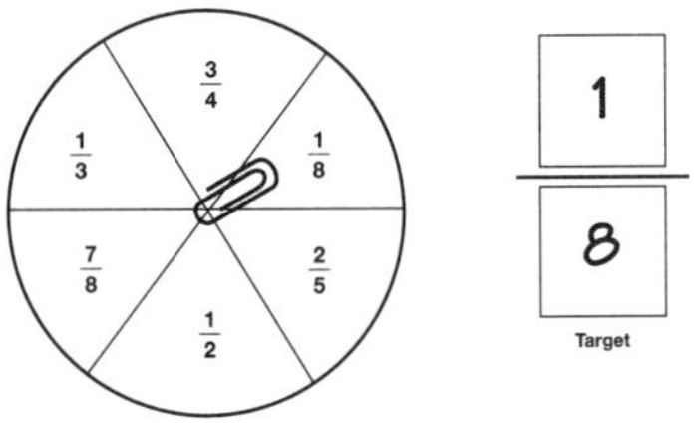

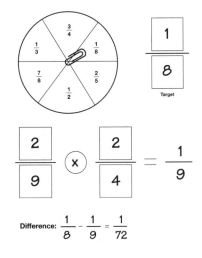

$$\frac{2}{9} \times \frac{2}{4} = \frac{1}{9}$$

Difference: $\frac{1}{8} - \frac{1}{9} = \frac{1}{72}$

1. Each student records the equation format shown at the bottom of the activity form. A volunteer student spins to generate a target fraction.
2. After five Digit Squares are randomly drawn and displayed, each student uses four of the five digits and any operation to try to equal the target fraction or an amount close to that fraction.
3. After students try different arrangements with their selected four digits, they record the one closest to the target fraction on their blank recording sheet.
4. Students are asked to find and record the difference between their solutions and the target fractions.
5. Different students share various solutions and identify how close they came to the target fraction.
6. In the next round another volunteer spins and identifies a new target fraction. Students use the same five displayed digits and repeat the process.
7. Future rounds can begin by drawing five new Digit Squares or spinning a new target fraction. Students find it advantageous to have repeat rounds with the same digit choices.

Making Connections

Promote reflection and make mathematical connections by asking:

- Which target fractions seemed easier to match? How might this be explained?
- How did using repeat digits make this activity easier or more difficult?

Target Fractions

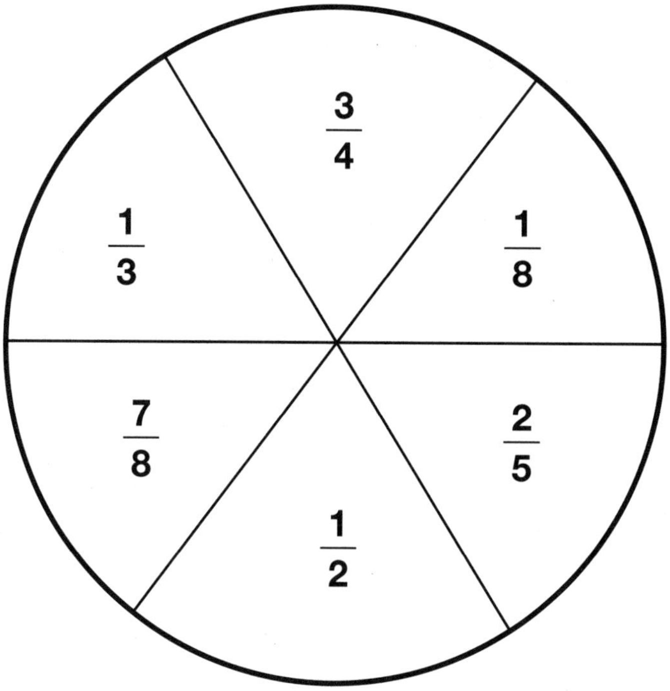

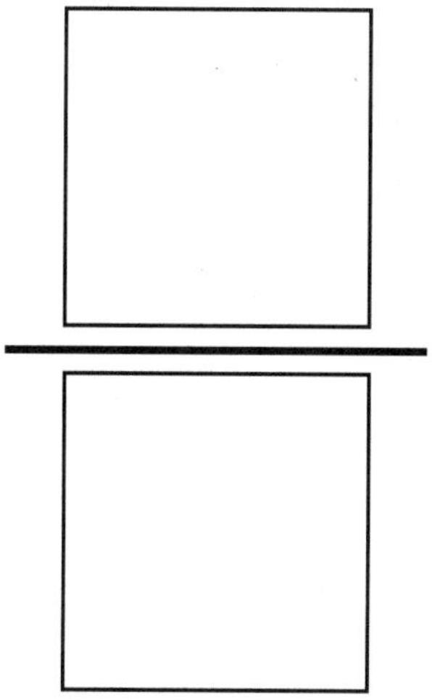

Target

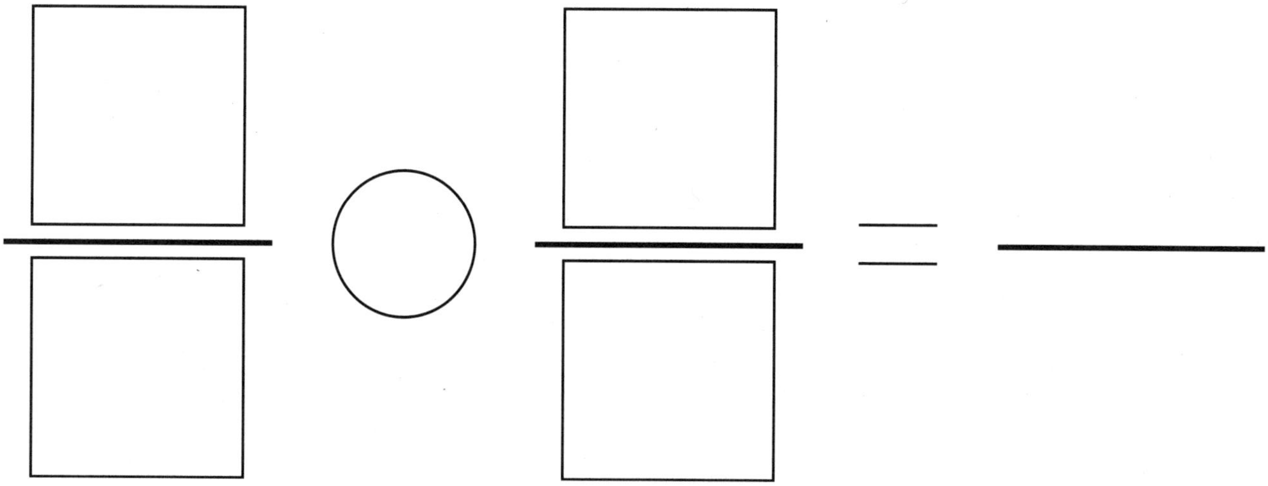

Difference: ___ − ___ = ___

Seeking Fractions

Tip *As students gain competence and confidence, include the "7" and "9" Digit Squares.*

Topic: Computing with Fractions

Object: Create qualifying fraction equations with all operations.

Groups: Whole class or small group

Materials

- *Seeking Fractions* recording sheet (for each participant), p. 77
- transparency of *Seeking Fractions* recording sheet
- set of Digit Squares (0, 7 & 9 removed), p. 151
- set of Digit Squares for each participant

Directions

1. The leader draws and displays four Digit Squares and the recording sheet.
2. Each student uses the same four drawn digits to create and record a fraction expression that completes the equation or inequality shown in Round 1.

   Example: 1, 3, 4, and 6 are drawn. A student records $\frac{1}{4} \times \frac{6}{3} = \frac{1}{2}$, while another player records $\frac{3}{6} + \frac{1}{4} < 1$.
3. Students are encouraged to seek additional qualifying expressions while waiting for all students to identify at least one.
4. The leader has students share different solutions for each format in Round 1, highlighting the variety of operations used. Students who found more than one solution are recognized.
5. The used Digit Squares are returned and four new Digit Squares are drawn and displayed for Round 2. During this round, each student attempts to use the displayed digits and a new operation to complete the equation or inequality. In order to create a second solution, a student is allowed to repeat an operation.
6. The leader has students share solutions and operations from their recording sheets.
7. In Round 3, the leader follows a similar process and encourages students to find solutions involving division or multiplication.

Making Connections

Promote reflection and make mathematical connections by asking:

- Which equations were easier to create? Why?
- What approach helped you find solutions requiring division?

Seeking Fractions

Recording Sheet

$$\frac{\square}{\square} \bigcirc \frac{\square}{\square}$$

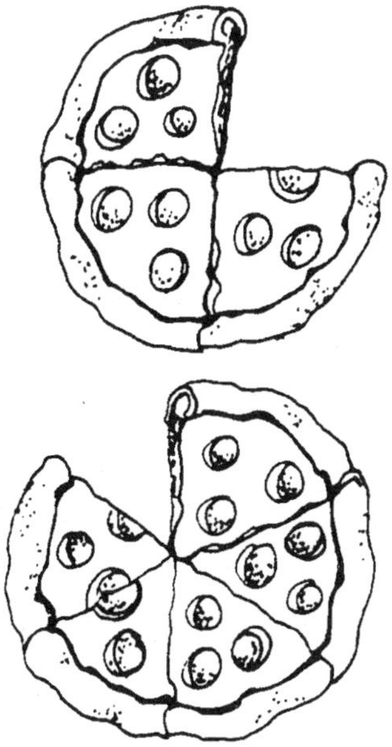

$\frac{3}{4}$

$\frac{5}{6}$

Round 1

$$\frac{\square}{\square} \bigcirc \frac{\square}{\square} = \frac{1}{2}$$

$$\frac{\square}{\square} \bigcirc \frac{\square}{\square} < 1$$

Round 2

$$\frac{\square}{\square} \bigcirc \frac{\square}{\square} = \underline{\qquad\qquad}$$

whole number

$$\frac{\square}{\square} \bigcirc \frac{\square}{\square} > 1$$

Round 3

$$\frac{\square}{\square} \bigcirc \frac{\square}{\square} = \underline{\qquad\qquad}$$

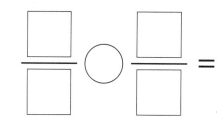

mixed number

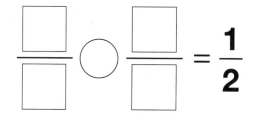

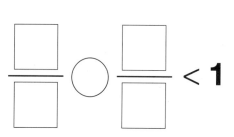

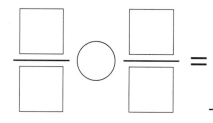

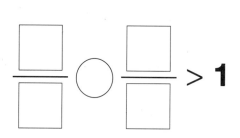

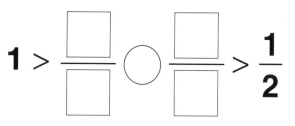

$$1 > \frac{\square}{\square} \bigcirc \frac{\square}{\square} > \frac{1}{2}$$

Date ____________ Name ____________

Partial Possibilities 1

STOP Don't start yet. Star problems that may have answers greater than 1. Express all fractions in lowest terms.

1. Order these fractions from smallest to largest. $\frac{4}{9}$ $\frac{1}{3}$ $\frac{1}{2}$ $\frac{5}{6}$ ____ ____ ____ ____

smallest largest

2. $\frac{2}{5} + \frac{3}{10} =$ ____

3. $\frac{5}{6} - \frac{1}{3} =$ ____

4. $\frac{1}{4} + \frac{2}{3} =$ ____

5. $1\frac{1}{5} + \frac{3}{10} =$ ____

6. $2 - \frac{3}{7} =$ ____

7. $4\frac{1}{4} - 2\frac{1}{2} =$ ____

8. $\frac{2}{3} - \frac{1}{2} =$ ____

9. $\frac{3}{4} \times \frac{2}{5} =$ ____

10. $8 \div \frac{1}{2} =$ ____

Go On What comes next? $2\frac{1}{4}$, $2\frac{5}{12}$, $2\frac{7}{12}$, $2\frac{3}{4}$, ____ ____ ____. Describe the pattern.

Date ____________ Name ____________

Partial Possibilities 2

STOP Don't start yet. Star problems that may have answers between 1 and 2. Express all fractions in lowest terms.

1. Order these fractions from smallest to largest. $\frac{1}{2}$ $\frac{5}{6}$ $\frac{3}{4}$ $\frac{2}{3}$ ____ ____ ____ ____

smallest largest

2. $\frac{4}{9} + \frac{2}{3} =$ ____

3. $\frac{1}{2} - \frac{5}{12} =$ ____

4. $\frac{2}{5} + \frac{1}{4} =$ ____

5. $2\frac{1}{6} + \frac{2}{3} =$ ____

6. $2 - \frac{5}{6} =$ ____

7. $3\frac{1}{2} - 1\frac{5}{12} =$ ____

8. $\frac{3}{4} - \frac{1}{6} =$ ____

9. $\frac{2}{3} \times \frac{1}{2} =$ ____

10. $\frac{5}{6} \div \frac{1}{2} =$ ____

Go On Name 3 different fractions between $\frac{1}{2}$ and $\frac{3}{4}$.

Date ____________ Name ____________

Partial Possibilities 3

STOP Don't start yet. Star problems that may have answers less than $\frac{1}{2}$.
Express all fractions in lowest terms.

1. Order these fractions from smallest to largest. $\frac{3}{4}$ $\frac{5}{6}$ $\frac{1}{3}$ $\frac{7}{12}$ ____ ____ ____ ____

smallest ... largest

2. $\frac{5}{12} + \frac{1}{6} =$ ____

3. $\frac{3}{4} - \frac{3}{8} =$ ____

4. $\frac{1}{3} + \frac{5}{8} =$ ____

5. $1\frac{3}{8} + \frac{1}{4} =$ ____

6. $2 - \frac{2}{5} =$ ____

7. $4\frac{5}{8} - 2\frac{3}{4} =$ ____

8. $\frac{3}{5} - \frac{1}{3} =$ ____

9. $\frac{3}{5} \times \frac{2}{3} =$ ____

10. $\frac{2}{3} \div \frac{1}{4} =$ ____

Go On Which doesn't fit? $\frac{4}{7}$, $\frac{8}{9}$, $\frac{3}{5}$, $\frac{6}{16}$, $\frac{5}{8}$. Describe your rule.

Date ____________ Name ____________

Partial Possibilities 4

STOP Don't start yet. Star problems that may have answers close to 1.
Express all fractions in lowest terms.

1. Order these fractions from smallest to largest. $\frac{1}{4}$ $\frac{1}{2}$ $\frac{2}{5}$ $\frac{3}{10}$ ____ ____ ____ ____

smallest ... largest

2. $\frac{5}{8} + \frac{1}{4} =$ ____

3. $\frac{2}{3} - \frac{7}{12} =$ ____

4. $\frac{3}{5} + \frac{1}{3} =$ ____

5. $2\frac{7}{12} + \frac{1}{6} =$ ____

6. $2 - \frac{3}{8} =$ ____

7. $3\frac{2}{5} - 1\frac{3}{10} =$ ____

8. $\frac{2}{3} - \frac{5}{8} =$ ____

9. $\frac{3}{4} \times \frac{1}{3} =$ ____

10. $7 \div \frac{1}{3} =$ ____

Go On What comes next? $1\frac{2}{3}$, $2\frac{2}{9}$, $2\frac{7}{9}$, $3\frac{1}{3}$, ____ ____ ____. Describe the pattern.

Date ____________ Name ____________

Partial Possibilities 5

STOP Don't start yet. Star problems that may have answers between $\frac{1}{2}$ and 1. Express all fractions in lowest terms.

1. Order these fractions from smallest to largest. $\frac{11}{12}$ $\frac{1}{2}$ $\frac{5}{6}$ $\frac{7}{8}$ _____ _____ _____ _____

smallest largest

2. $\frac{3}{4} + \frac{1}{12} =$ _____

3. $\frac{1}{2} - \frac{1}{6} =$ _____

4. $\frac{3}{4} + \frac{1}{6} =$ _____

5. $1\frac{5}{9} + \frac{1}{3} =$ _____

6. $2 - \frac{4}{9} =$ _____

7. $4\frac{1}{6} - 2\frac{1}{3} =$ _____

8. $\frac{7}{8} - \frac{1}{6} =$ _____

9. $\frac{2}{3} \times \frac{3}{4} =$ _____

10. $\frac{3}{4} \div \frac{1}{4} =$ _____

Go On Name three different fractions between $\frac{1}{3}$ and $\frac{2}{3}$.

Date ____________ Name ____________

Partial Possibilities 6

STOP Don't start yet. Star problems that may have answers that won't need to be reduced to lowest terms. Express all fractions in lowest terms.

1. Order these fractions from smallest to largest. $\frac{3}{8}$ $\frac{1}{3}$ $\frac{1}{4}$ $\frac{5}{6}$ _____ _____ _____ _____

smallest largest

2. $\frac{5}{12} + \frac{3}{4} =$ _____

3. $\frac{11}{12} - \frac{2}{3} =$ _____

4. $\frac{1}{3} + \frac{3}{8} =$ _____

5. $2\frac{5}{12} + \frac{1}{4} =$ _____

6. $2 - \frac{7}{10} =$ _____

7. $3\frac{1}{6} - 1\frac{5}{12} =$ _____

8. $\frac{4}{5} - \frac{2}{3} =$ _____

9. $\frac{5}{6} \times \frac{2}{3} =$ _____

10. $\frac{4}{5} \div \frac{1}{3} =$ _____

Go On What else belongs? $\frac{11}{12}$, $\frac{4}{5}$, $\frac{7}{8}$, $\frac{3}{4}$. Describe your rule.

Ten Tallies Win

Topic: Reducing and Comparing Fractions

Object: Be the first to generate ten larger proper fractions.

Groups: 2 players

Materials for each pair

- *Ten Tallies Win* recording sheet, p. 82
- 8 number cubes (1–6)
- scratch paper for each player

Directions

1. Each player rolls four number cubes to generate four numbers.
2. Each player attempts to get the largest possible proper fraction by multiplying two of their rolled numbers to generate a numerator and the other two rolled numbers to produce a denominator. If appropriate, generated fractions are reduced to lowest terms.

Numbers Rolled	Fraction	Reduced Fraction
6 5 3 2	$\frac{12}{15}$	$\frac{4}{5}$
1 3 3 4	$\frac{4}{9}$	$\frac{4}{9}$

3. After each player explains how she or he produced the generated fraction, the players determine which fraction is larger. The player with the larger fraction must adequately justify that her or his fraction is larger.
4. The player with the larger fraction and accepted justification records a tally mark.
5. These steps are repeated until one player wins by recording ten tally marks.

__Tips__ Increase success rate by allowing players to each roll five number cubes and select only four of the five cubes for each turn. An interesting variation is awarding the player with the smaller ***improper fraction*** *with a tally mark. (This variation works well using four of the five rolled number cubes.)*

Making Connections

Promote reflection and make mathematical connections by asking:

- What approaches were effective to prove which fraction was larger?
- What number arrangements generated larger proper fractions? Explain.

Ten Tallies Win

Recording Sheet

Ten Tallies Win

Numbers Rolled	Fraction	Reduced Fraction
— — — —	—	—
— — — —	—	—
— — — —	—	—
— — — —	—	—
— — — —	—	—
— — — —	—	—
— — — —	—	—
— — — —	—	—
— — — —	—	—
— — — —	—	—
— — — —	—	—
— — — —	—	—
— — — —	—	—
— — — —	—	—
— — — —	—	—

Tally

Ten Tallies Win

Numbers Rolled	Fraction	Reduced Fraction
— — — —	—	—
— — — —	—	—
— — — —	—	—
— — — —	—	—
— — — —	—	—
— — — —	—	—
— — — —	—	—
— — — —	—	—
— — — —	—	—
— — — —	—	—
— — — —	—	—
— — — —	—	—
— — — —	—	—
— — — —	—	—
— — — —	—	—

Tally

Ordered Fractions

Topic: Comparing Fractions

Object: Record fractions in an ascending order.

Groups: 2 players or pair players

Materials for each group

- *Ordered Fractions* recording sheet (for each player), p. 84
- 2 sets of Digit Cards (0 removed), p. 150

Tip *Prepare students for independent success with this game by dividing the class in half and playing this as a team game, with each team publicly reaching decisions and displaying results on separate recording sheets.*

Directions

1. One player mixes the two sets of Digit Cards and stacks them face down. Each player draws two cards and uses the drawn digits to form a fraction between 0 and 3.

   Example: If 7 and 2 are drawn, the player would choose $\frac{2}{7}$ since $\frac{7}{2}$ does not fit within the given range.
2. Keeping in mind the relative value positions of fractions between 0 and 3, each player records his or her choice in one of the cells along the pathway. After recording their choices, the players share their decisions with each other. If players accept their opponent's displayed order, play continues. Once a fraction is recorded, it cannot be moved. Drawn cards are set aside to be used later.
3. Players draw two new Digit Cards and repeat these same steps. The recording of an equivalent fraction in an adjacent cell is allowed.
4. When the stack of Digit Cards gets low, all Digit Cards are mixed and restacked.
5. A player loses a turn if the drawn digit cannot form a fraction that can be placed in any of the remaining cells.
6. Play continues until one player correctly completes a pathway that orders fractions from smallest to largest.

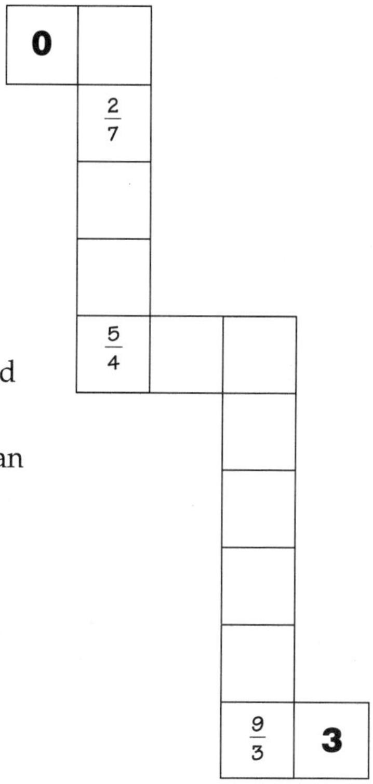

Making Connections

Promote reflection and make mathematical connections by asking:

- How did you decide where to place your fractions?
- Which fractions were more difficult to place? Please explain.

Ordered Fractions

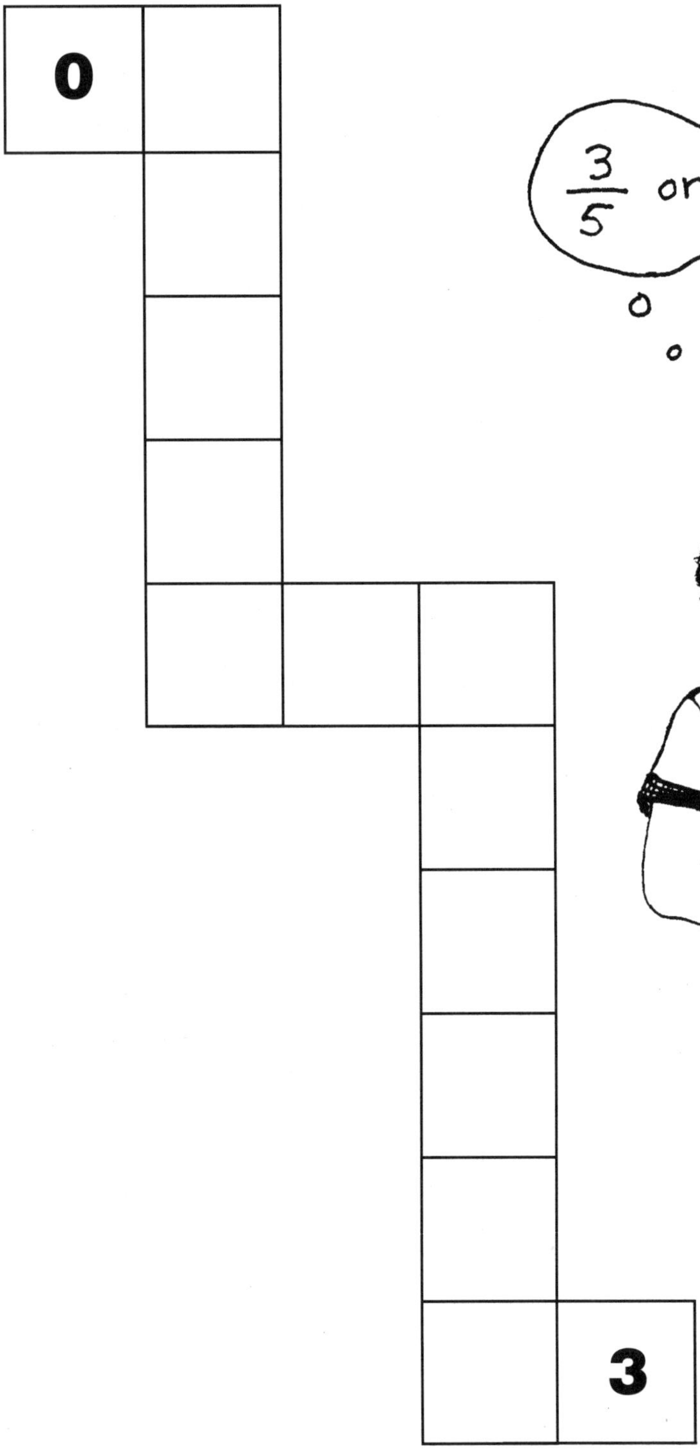

Fraction Arrangements

Topic: Mentally Computing Fractions

Object: Compute two created fractions to reach close to a target amount.

Groups: Pair players or 2 players

Materials for each group

- *Fraction Arrangements* recording sheet (for each player), p. 86
- 3 sets of Digit Cards (0's and 7's removed), p. 150

Tip Use Fraction Arrangements B, *p. 87, when students become comfortable with* Fraction Arrangements A *and are ready for the greater challenge of mentally multiplying and dividing fractions.*

Directions

1. One player mixes the three sets of Digit Cards and stacks them face down. For Round 1 each pair draws five cards and uses four of the drawn digits to form two fractions that equal a sum close to one whole.
2. When both pairs are ready, they share their fraction arrangements, announcing the resulting sum. After the opposing pair verifies the arrangement as correct, the equation is recorded.
3. Next members of each pair determine who is closest to the target number by computing the difference between their answer and the target amount. After the opposing pairs verify these differences, the pair with the lowest difference receive 1 point for that round.
4. The used cards are set aside and each pair draws five new cards. The pairs play round two by following the same steps, while seeking a difference close to $\frac{1}{2}$.
5. After every two rounds remix and restack all the Digit Cards.
6. In the last round, the pairs get to choose which operation to use with their two arranged fractions.
7. After up to eight rounds are played, the scores are totaled. The pair with the highest score wins.

		Answer	Target	Difference	Score
Round 1	$\frac{3}{6} + \frac{6}{9} =$	$1\frac{1}{6}$	1	$\frac{1}{6}$	1
Round 2	$\frac{5}{4} - \frac{6}{8} =$	$\frac{1}{2}$	$\frac{1}{2}$	0	

Making Connections

Promote reflection and make mathematical connections by asking:

- Which fractions were easier to reach? Explain.
- Which digits were more difficult to place?
- What arrangements helped you get close to the target numbers?

Fraction Arrangements A

Round		Answer	Target	Difference	Score
Round 1	$\frac{\square}{\square} + \frac{\square}{\square} =$	___	1		
Round 2	$\frac{\square}{\square} - \frac{\square}{\square} =$	___	$\frac{1}{2}$		
Round 3	$\frac{\square}{\square} - \frac{\square}{\square} =$	___	$\frac{1}{4}$		
Round 4	$\frac{\square}{\square} + \frac{\square}{\square} =$	___	$\frac{3}{4}$		
Round 5	$\frac{\square}{\square} - \frac{\square}{\square} =$	___	1		
Round 6	$\frac{\square}{\square} + \frac{\square}{\square} =$	___	$1\frac{1}{4}$		
Round 7	$\frac{\square}{\square} - \frac{\square}{\square} =$	___	$\frac{3}{4}$		
Round 8	$\frac{\square}{\square} \bigcirc \frac{\square}{\square} =$	___	$\frac{2}{3}$		

______ **Total Score**

Fraction Arrangements B

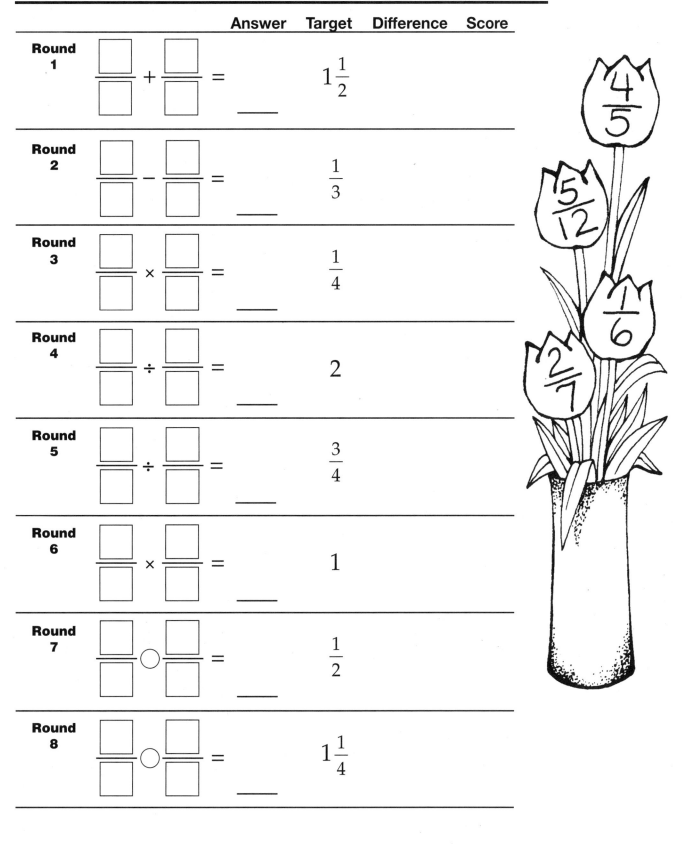

		Answer	Target	Difference	Score
Round 1	$\frac{\square}{\square} + \frac{\square}{\square} =$	____	$1\frac{1}{2}$		
Round 2	$\frac{\square}{\square} - \frac{\square}{\square} =$	____	$\frac{1}{3}$		
Round 3	$\frac{\square}{\square} \times \frac{\square}{\square} =$	____	$\frac{1}{4}$		
Round 4	$\frac{\square}{\square} \div \frac{\square}{\square} =$	____	2		
Round 5	$\frac{\square}{\square} \div \frac{\square}{\square} =$	____	$\frac{3}{4}$		
Round 6	$\frac{\square}{\square} \times \frac{\square}{\square} =$	____	1		
Round 7	$\frac{\square}{\square} \bigcirc \frac{\square}{\square} =$	____	$\frac{1}{2}$		
Round 8	$\frac{\square}{\square} \bigcirc \frac{\square}{\square} =$	____	$1\frac{1}{4}$		

____
Total Score

Date ____________ Name ____________________

If... Then...

Solve each of the following problems. The first is done for you.

If ▲ = $\frac{1}{2}$, then . . .

1. ▲▲ = 1

2. ▲▲▲▲▲ = _____

3. ▲▲▲▲▲▲▲▲▲▲ = _____

4. ▲▲▲▲ = _____

If ★★★ = $1\frac{1}{2}$, then . . .

5. ★★ = _____

6. ★★★★ = _____

7. ★★★★★★ = _____

8. ★ = _____

If ✿✿ = 3, then . . .

9. ✿✿✿✿ = _____

10. ✿ = _____

11. ✿✿✿✿✿ = _____

12. ✿✿✿ = _____

If ■■ = $\frac{1}{4}$, then . . .

13. ■ = _____

14. ■■■■■■■■ = _____

15. ■■■■■ = _____

16. ■■■■■■■■■■■■ = _____

If ❤❤❤❤❤ = 10, then . . .

17. ❤❤ = _____

18. ❤❤❤❤❤❤ = _____

19. ❤❤❤ = _____

20. ❤ = _____

If ○○○ = $\frac{1}{2}$, then . . .

21. ○ = _____

22. ○○ = _____

23. ○○○○○○○○ = _____

24. ○○○○○ = _____

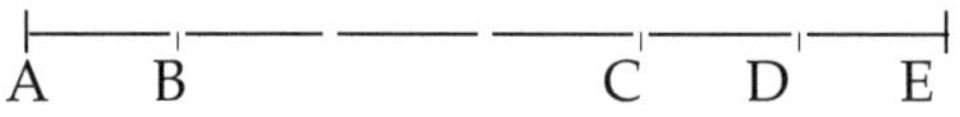

25. If A = 0 and E = 1, what's the value of B? ______ C? _____ D? _____

26. If A = 0 and E = $1\frac{1}{2}$, what's the value of B? ______ C? _____ D? _____

27. ***If A = 0 and E = 4, what's the value of B? ______ C? _____ D? _____

Date ____________________ Name ________________________________

Digits to Fractions I

Use each of these displayed digits once in each equation to create fractions which make each equation true.

Use **1** **2** **4** **5**

1. $\frac{\square}{\square} + \frac{\square}{\square} = \frac{13}{20}$
2. $\frac{\square}{\square} - \frac{\square}{\square} = \frac{3}{4}$
3. $\frac{\square}{\square} + \frac{\square}{\square} = 1\frac{3}{10}$
4. $\frac{\square}{\square} + \frac{\square}{\square} = \frac{7}{10}$
5. $\frac{\square}{\square} - \frac{\square}{\square} = \frac{3}{10}$
6. $\frac{\square}{\square} - \frac{\square}{\square} = 2\frac{1}{4}$

Use **2** **3** **5** **8**

7. $\frac{\square}{\square} + \frac{\square}{\square} = \frac{17}{20}$
8. $\frac{\square}{\square} - \frac{\square}{\square} = \frac{14}{15}$
9. $\frac{\square}{\square} + \frac{\square}{\square} = 2\frac{4}{15}$
10. $\frac{\square}{\square} + \frac{\square}{\square} = 1\frac{7}{24}$
11. $\frac{\square}{\square} - \frac{\square}{\square} = \frac{7}{20}$
12. $\frac{\square}{\square} - \frac{\square}{\square} = 2\frac{1}{3}$

Date ______________ Name ______________________

Digits to Fractions II

Use each of these displayed digits once in each equation to create fractions which make each equation true.

Use **1** **2** **3** **4**

1. $\frac{\square}{\square} \times \frac{\square}{\square} = \frac{1}{6}$

2. $\frac{\square}{\square} \times \frac{\square}{\square} = \frac{3}{8}$

3. $\frac{\square}{\square} \div \frac{\square}{\square} = 6$

4. $\frac{\square}{\square} \times \frac{\square}{\square} = \frac{2}{3}$

5. $\frac{\square}{\square} \div \frac{\square}{\square} = 1\frac{1}{2}$

6. $\frac{\square}{\square} \div \frac{\square}{\square} = 2\frac{2}{3}$

Use **1** **4** **5** **6**

7. $\frac{\square}{\square} \times \frac{\square}{\square} = \frac{3}{10}$

8. $\frac{\square}{\square} \times \frac{\square}{\square} = \frac{5}{24}$

9. $\frac{\square}{\square} \div \frac{\square}{\square} = 7\frac{1}{2}$

10. $\frac{\square}{\square} \times \frac{\square}{\square} = \frac{2}{15}$

11. $\frac{\square}{\square} \div \frac{\square}{\square} = 4\frac{4}{5}$

12. $\frac{\square}{\square} \div \frac{\square}{\square} = 3\frac{1}{3}$

Date ____________________ Name ________________________________

Fraction Choices I

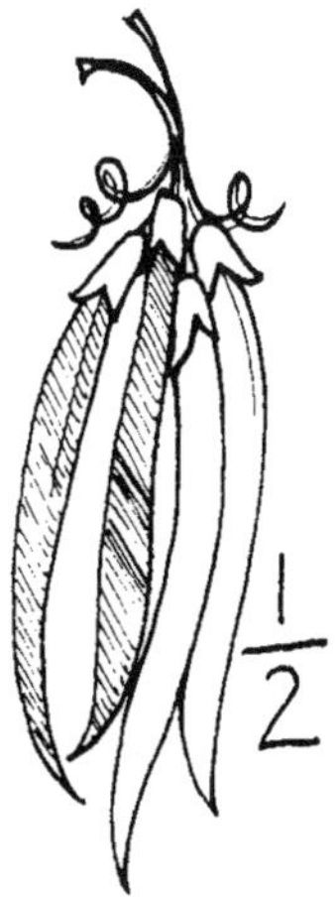

Use three of these four fractions to make each equation true.

$\frac{1}{2}$	$\frac{3}{8}$	$\frac{1}{8}$	$\frac{1}{4}$

1. $\square + \square + \square = \frac{3}{4}$

2. $\square + \square - \square = \frac{5}{8}$

3. $\square - \square + \square = \frac{3}{8}$

4. $\square - \square - \square = 0$

5. $\square + \square + \square = 1\frac{1}{8}$

6. $\square - \square + \square = \frac{1}{4}$

7. $\square - \square - \square = \frac{1}{8}$

8. $\square + \square - \square = 0$

Use three of these four fractions to make each equation true.

$\frac{1}{3}$	$\frac{2}{3}$	$\frac{1}{2}$	$\frac{1}{6}$

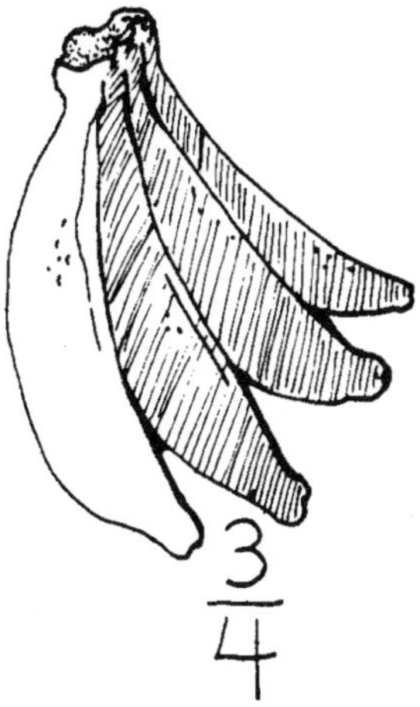

9. $\square + \square + \square = 1\frac{1}{2}$

10. $\square + \square - \square = 1$

11. $\square - \square - \square = 0$

12. $\square - \square + \square = \frac{1}{3}$

13. $\square + \square + \square = 1\frac{1}{3}$

14. $\square - \square - \square = \frac{1}{6}$

15. $\square + \square - \square = \frac{5}{6}$

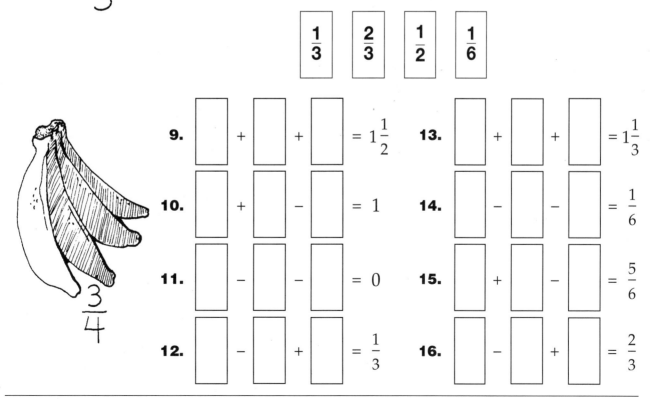

16. $\square - \square + \square = \frac{2}{3}$

Date ____________ Name ____________________

Fraction Choices II

Use three of these four fractions to make each equation true.

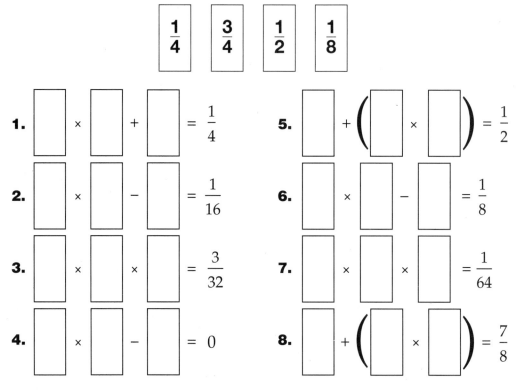

$\frac{1}{4}$ $\frac{3}{4}$ $\frac{1}{2}$ $\frac{1}{8}$

1. $\square \times \square + \square = \frac{1}{4}$

2. $\square \times \square - \square = \frac{1}{16}$

3. $\square \times \square \times \square = \frac{3}{32}$

4. $\square \times \square - \square = 0$

5. $\square + \left(\square \times \square\right) = \frac{1}{2}$

6. $\square \times \square - \square = \frac{1}{8}$

7. $\square \times \square \times \square = \frac{1}{64}$

8. $\square + \left(\square \times \square\right) = \frac{7}{8}$

Use three of these four fractions to make each equation true.

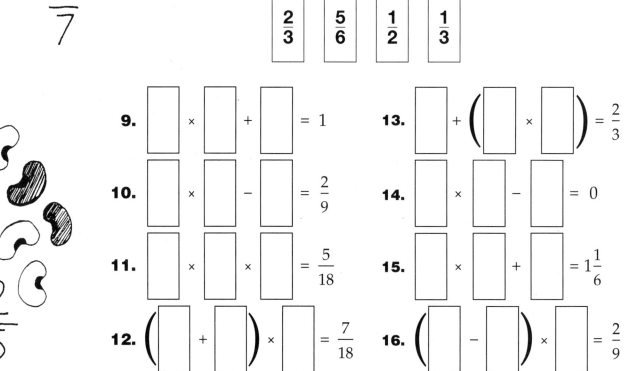

$\frac{2}{3}$ $\frac{5}{6}$ $\frac{1}{2}$ $\frac{1}{3}$

9. $\square \times \square + \square = 1$

10. $\square \times \square - \square = \frac{2}{9}$

11. $\square \times \square \times \square = \frac{5}{18}$

12. $\left(\square + \square\right) \times \square = \frac{7}{18}$

13. $\square + \left(\square \times \square\right) = \frac{2}{3}$

14. $\square \times \square - \square = 0$

15. $\square \times \square + \square = 1\frac{1}{6}$

16. $\left(\square - \square\right) \times \square = \frac{2}{9}$

Decimals

Assumptions Decimal operations have previously been taught and reviewed, emphasizing understanding and building decimal sense. Concrete objects and visual models, such as coins or base 10 materials and grids, have been used extensively.

Section Overview and Suggestions

Sponges

Where's the Point? pp. 94–95

Finding Decimal Differences p. 96

Decimal Line-ups pp. 97–98

Going for More pp. 99–100

These whole-class or small-group warm-ups are repeatable. They help build decimal sense as students enhance their abilities to mentally compute decimals. Repeated use of all sponges, especially *Decimal Line-ups* and *Going for More,* will ensure greater success with *Neighbors Count* and the Independent Activities.

Skill Checks

Partial Possibilities 7–12 pp. 101–103

The Skill Checks provide a way for parents, students, and you to see students' improvement with decimal conversion and computation. Copies can be cut in half so that each check may be used at a different time. Remember to have all students respond to the STOP, number sense task, before solving the ten problems.

Games

Rolling Decimals pp. 104–105

Totaling One pp. 106–107

Neighbors Count pp. 108–109

These open-ended and repeatable Games actively involve students in converting or computing decimals, while students enhance their strategic thinking abilities. The challenging, yet popular *Neighbors Count* game is also ideal for use at home.

Independent Activities

Identifying Missing Decimals p. 110

Fitting Decimals pp. 111–112

These activities easily engage students in practicing decimal operations independently. These activities require extensive mental computation, and lead to improved decimal sense. It is highly recommended that students complete and correct the two sections of *Identifying Missing Decimals* at different times. *Fitting Decimals* includes two versions, allowing practice of addition and subtraction, or multiplication and division. Either of these Independent Activities can be replicated with new numbers to easily provide additional practice.

Where's the Point?

Topic: Decimal Relationships Resulting from Computation

Object: Correctly place the decimal points in decimals when their decimal sum or product is given.

Groups: Whole class or small group

Materials

- transparency of *Where's the Point?* activity form, p. 95
- additional prepared problems (*optional*)

Tip If an easier version is desired, introduce this warm-up with a simpler addition problem, like **6 2 5 + 4 9 7 = 6 7.4 7** *or a subtraction example, like* **3 1 2 8 – 7 6 = 3 0 . 5 2.**

Directions

1. The leader displays only the first problem line of the *Where's the Point?* activity form and asks, "If the given sum is correct, where should the decimal points be placed in the addends?"
2. Students signal to the leader when they have an answer. A selected student then names the three decimal addends by correctly reading the decimal sentence.

   Example: For the first problem the responding student would state, "Three and six tenths plus forty-five and twenty-six hundredths plus nine and forty-four hundredths equals fifty-eight and three tenths."

   3.6 + 4 5.2 6 + 9.4 4 = 5 8.3

   5 5.7 + 9.6 + 3 4.2 3 = 9 9.5 3
3. After the leader displays the next problem and provides adequate time for most students to signal readiness to respond, another student identifies the three decimal addends and total. This process continues through the first five problems.
4. The leader explains that the sixth problem involves multiplication of two decimal factors. Again, the students signal when they know where to place the decimal points in the factors to yield the given product.
5. At this point it is important for the leader to ask, "Did anyone solve the problem a different way?" (For the sixth problem there are four possible solutions.)
6. This modified procedure continues through the remaining multiplication rows. (For additional practice, students can be paired up and create problems for classmates to solve.)

Making Connections

Promote reflection and make mathematical connections by asking:

- What approach helped you decide where to place the decimal points?
- Why do the multiplication problems allow multiple arrangements of the decimal points? Please explain.

Where's the Point?

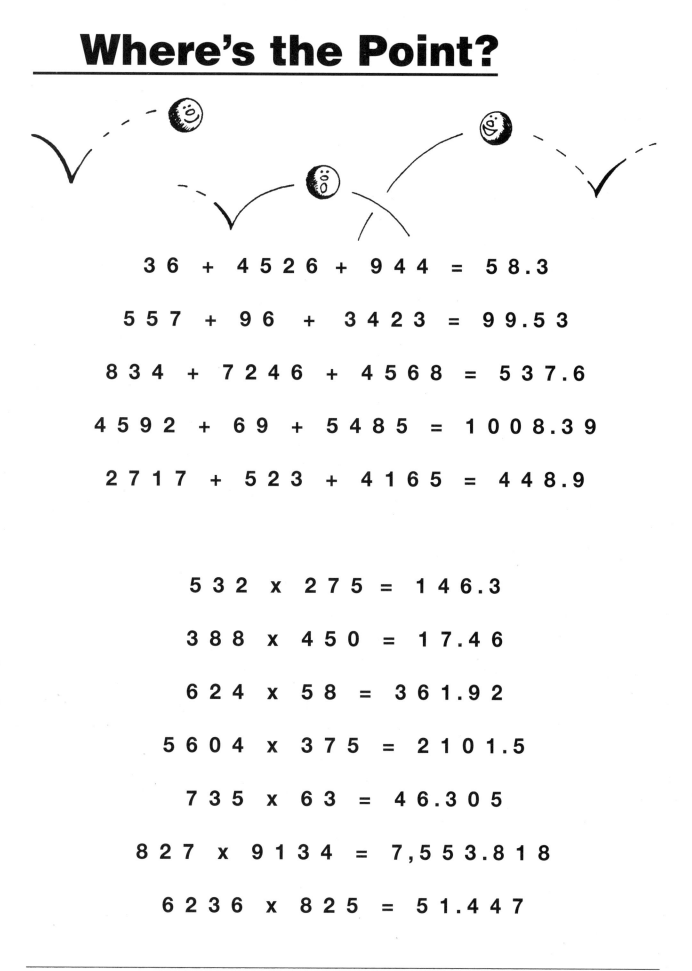

36 + 4526 + 944 = 58.3

557 + 96 + 3423 = 99.53

834 + 7246 + 4568 = 537.6

4592 + 69 + 5485 = 1008.39

2717 + 523 + 4165 = 448.9

532 x 275 = 146.3

388 x 450 = 17.46

624 x 58 = 361.92

5604 x 375 = 2101.5

735 x 63 = 46.305

827 x 9134 = 7,553.818

6236 x 825 = 51.447

Finding Decimal Differences

Topic: Mental Subtraction of Decimals

Object: Identify two decimals whose difference equals a given amount.

Groups: Whole class or small group

Materials

- chalkboard or overhead projector

Tips *Increase the level of difficulty by expanding the list to more than six choices and including more challenging decimals.*

Directions

1. The leader displays a group of numbers. (See first boxed list.)
2. The leader familiarizes students with the displayed numbers by asking them to ...

   Find two numbers whose difference is greater than 40.

   ... any others?

   Find two numbers whose difference is less than 20.

   ... any others?

   Find two numbers whose difference is less than 10.

   ... any others?

   Find two numbers with a difference ending in .3.

   Find two numbers with a difference close to 50.

8.6	55
23	63.4
30.7	81

3. Next the leader announces an exact possible difference, such as 22.1. The students try to identify two listed numbers whose difference is 22.1.
4. The leader continues stating differences and the students respond with solutions.

   Other possible differences:
   50.3, 24.3, 40.4, 58, 46.4, 8.4, 14.4, 17.6, 7.7, 72.4, 32.7, 54.8, 26

5. Provide students with a more challenging listing. (See second boxed list.)

   Possible differences for the second group:
   49.5, 20.8, 6.95, 33.6, 21.75, 30.56, 85.91, 28.7, 23.61, 12.8, 55.35, 52.31, 42.55, 62.3, 73.11

.59	52.9
24.2	73.7
31.15	86.5

Making Connections

Promote reflection and make mathematical connections by asking:

- What helped you quickly locate differences?
- What patterns did you notice?

Decimal Line-ups

Topic: Mental Multiplication of Decimals

Object: Cover four decimals in a row.

Groups: Whole class or small group

Materials

- transparency of *Decimal Line-ups* activity board, p. 98
- markers (different kind for each team)
- calculator

Tips If time is short, allow three in a row for a win. This sponge can be easily adapted into a game for two players or pair players.

Directions

1. Divide the class or group into two teams. Each team uses a different kind of marker.
2. A member from the first team selects and announces two factors (one from each row in the "Choices" box). After choices are announced, students are allowed to use a calculator or pencil and paper to find the product. The first team covers the resulting product on the activity board with an appropriately colored marker.
3. A member of the other team selects and announces two factors whose product can be found and then be checked by a calculator or paper and pencil. If the resulting product is not covered, it is covered by that team's marker.
4. Teams continue alternating turns, announcing factors, and finding products.
5. The first team to have four markers in a row horizontally, vertically, or diagonally wins.

Choices			
1.3	2.8	4.5	6.2
29	34	57	83

130.5	81.2	353.4	44.2
107.9	210.8	153	232.4
159.6	256.5	37.7	179.8
514.6	74.1	373.5	95.2

Making Connections

Promote reflection and make mathematical connections by asking:

- What strategy helped you identify the desired factors?
- How could this warm-up be modified for continued use?

Decimal Line-ups

Choices			
1.3	2.8	4.5	6.2
29	34	57	83

130.5	81.2	353.4	44.2
107.9	210.8	153	232.4
159.6	256.5	37.7	179.8
514.6	74.1	373.5	95.2

Going For More

Topic: Mentally Computing Decimals

Object: Thoughtfully place digits to generate a high total.

Groups: Whole class or small group

Materials

- *Going for More* recording sheet (for each participant), p. 100
- 2 sets of transparent Digit Squares (0 removed), p. 151

Tips Students benefit from collaborating as pairs when the activity is first introduced. After students gain confidence with this version, have them aim for low scores.

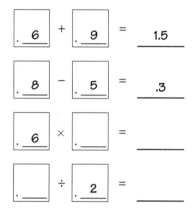

Directions

1. After the leader draws and displays one Digit Square, each student independently records the displayed digit in one of the eight boxes in the equations on the recording sheet.
2. When it is determined that each student has recorded the displayed digit, a "new" digit is drawn. (Drawn digits are not returned to the container until the end of the round.)
3. It is important for students to realize that in the second row they must subtract a smaller number from a larger number and that in the fourth row they should round to the nearest hundredth when finding the quotient.
4. After eight digits are displayed and recorded, the students compute the resulting four expressions and then total the sum, difference, product, and quotient.
5. Students next rearrange the same eight digits in the lower half of their recording sheet with the goal of generating a higher total.
6. The leader asks students to share their results for both arrangements.
7. If time allows, conduct an additional round of this sponge.

.6	+	.9	=	1.5
.8	−	.5	=	.3
.6	×	.	=	
.	÷	.2	=	
				Total

Making Connections

Promote reflection and make mathematical connections by asking:

- Where is it best to place higher digits?
- What will you do differently next time to generate a high total?

Going for More

.___	+	.___	=	______
.___	−	.___	=	______
.___	×	.___	=	______
.___	÷	.___	=	______

______
Total

Going for even more with the same digits

.____ + .____ = ______

.____ − .____ = ______

.____ × .____ = ______

.____ ÷ .____ = ______

______
Total

Going for More

.___	+	.___	=	______
.___	−	.___	=	______
.___	×	.___	=	______
.___	÷	.___	=	______

______
Total

Going for even more with the same digits

.____ + .____ = ______

.____ − .____ = ______

.____ × .____ = ______

.____ ÷ .____ = ______

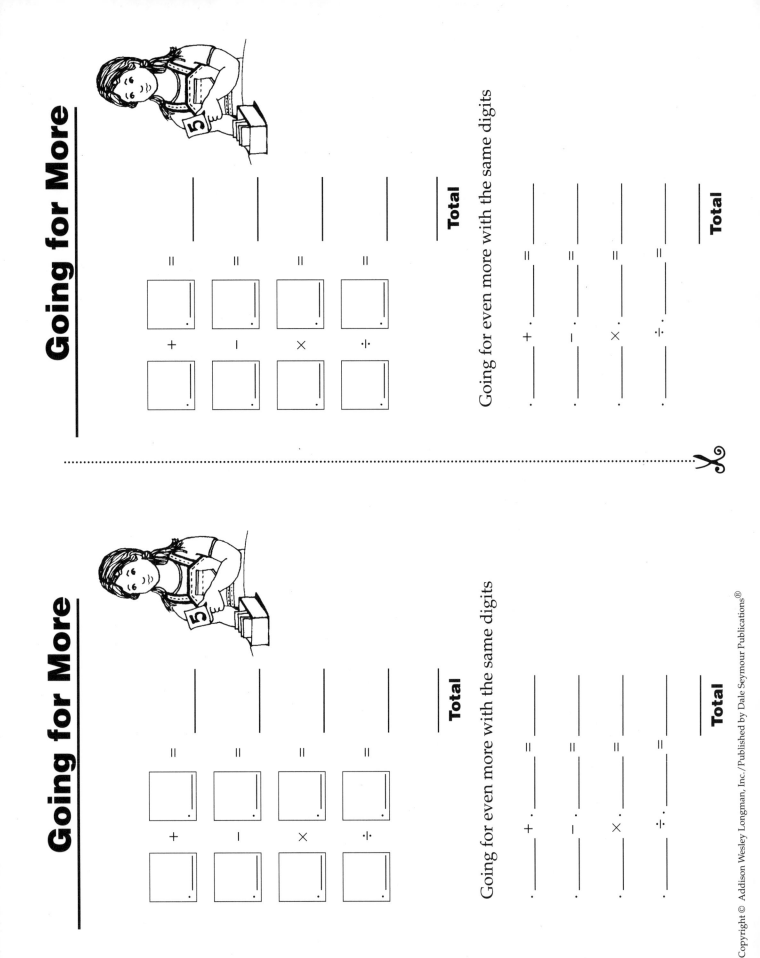

______
Total

Date ____________ Name ____________

Partial Possibilities 7

STOP Don't start yet. Star a problem that may have the largest answer.

1. Write the number sixteen and four-tenths. ________

2. Change $\frac{3}{8}$ to a decimal. ________

3. Order from least to greatest. $\frac{2}{3}, \frac{1}{2}, 0.7$ ____________

4. Place the decimal point in each addend. 6 2 3 + 5 8 7 + 5 0 7 = 70

5. 34.7 + 6.75 = ________

6. 41.2 − 1.36 = ________

7. 0.36 + 0.47 + 0. ____ = 1

8. 1.3 × 7 = ________

9. 2.7 × 1.4 = ________

10. 7.2 ÷ 4 = ________

Go On What comes next? 3.82, 6.42, 9.02, 11.62, ________, ________, ________
Describe the pattern.

Date ____________ Name ____________

Partial Possibilities 8

STOP Don't start yet. Star a problem that may have the smallest answer.

1. Write the number five and thirty-six hundredths. ________

2. Change $\frac{4}{5}$ to a decimal. ________

3. Order from least to greatest. $\frac{1}{3}, 0.18, \frac{1}{5}$ ____________

4. Place the decimal point in each addend. 4 1 9 + 4 4 + 2 7 6 6 = 70

5. 9.26 + 0.89 = ________

6. 2.3 − 0.58 = ________

7. 0.45 + 0.29 + 0. ____ = 1

8. 0.5 × 8 = ________

9. 3.2 × 0.5 = ________

10. 69 ÷ 0.3 = ________

Go On Use the digits 4, 5, 6, 8, and 9 and two decimal points to write an addition problem with the largest possible sum. Describe your strategy.

Date ____________ Name ____________________

Partial Possibilities 9

STOP Don't start yet. Star problems that may have answers in the hundredths.

1. Write the number eleven and eight-tenths. ________

2. Change $\frac{7}{8}$ to a decimal. ________

3. Order from least to greatest. $\frac{1}{2}, \frac{2}{5}, 0.51$ ______________

4. Place the decimal point in each addend. 7 1 6 + 4 2 5 + 1 5 3 4 = 65

5. $62.17 + 8.9 =$ ________

6. $30.12 - 7.55 =$ ________

7. $0.19 + 0.37 + 0.$____ $= 1$

8. $2.4 \times 3 =$ ________

9. $0.5 \times .36 =$ ________

10. $1.5 \div 3 =$ ________

Go On Use 4, 6, 7, and 8 and decimal point(s) to complete this equation.

☐ ☐ × ☐ ☐ = 35.88

Date ____________ Name ____________________

Partial Possibilities 10

STOP Don't start yet. Star a problem that may have an answer greater than 10.

1. Write the number seven and twenty-one hundredths. ________

2. Change $\frac{2}{9}$ to a decimal. ________

3. Order from least to greatest. $\frac{1}{3}$, 0.4, 0.37 ______________

4. Place the decimal point in each addend. 3 2 0 8 + 8 6 2 + 7 9 3 = 120

5. $5.8 + 27.6 =$ ________

6. $5.6 - 2.84 =$ ________

7. $0.08 + 0.64 + 0.$ ____ $= 1$

8. $0.6 \times 4 =$ ________

9. $4.5 \times 0.03 =$ ________

10. $2.4 \div 6 =$ ________

Go On What comes next? 4.3, 7.9, 11.5, 15.1, ________, ________, ________

Describe the pattern.

Date ____________ Name ____________________

Partial Possibilities 11

STOP Don't start yet. Star a problem that may have an answer in the tenths.

1. Write the number nine and seven hundredths. ________

2. Change $\frac{5}{8}$ to a decimal. ________

3. Order from least to greatest. $0.5, \frac{2}{3}, \frac{3}{8}$ ____________

4. Place the decimal point in each addend. 1 2 0 4 + 3 9 7 + 2 5 6 3 = 150

5. $7.38 + 46.5 =$ ________ **6.** $1.1 - 0.7 =$ ________ **7.** $0.23 + 0.47 + 0.$ ____ $= 1$

8. $20 \times 0.6 =$ ________ **9.** $0.5 \times 0.21 =$ ________ **10.** $12 \div 0.4 =$ ________

Go On Use the digits 2, 3, 4, 6, and 7 and two decimal points to write a subtraction problem with the smallest possible difference. Describe your strategy.

Date ____________ Name ____________________

Partial Possibilities 12

STOP Don't start yet. Star problems that may have answers less than 5.

1. Write the number twelve and two-hundredths. ________

2. Change $\frac{5}{6}$ to a decimal. ________

3. Order from least to greatest. $\frac{5}{6}, \frac{1}{2}, 0.72$ ____________

4. Place the decimal point in each addend. 5 2 7 + 4 0 8 1 + 6 4 9 = 100

5. $4.97 + 0.36 =$ ________ **6.** $10.4 - 5.87 =$ ________ **7.** $0.36 + 0.29 + 0.$ ____ $= 1$

8. $2.7 \times 4 =$ ________ **9.** $0.46 \times 2.2 =$ ________ **10.** $3 \div 0.5 =$ ________

Go On Use the digits 3, 5, 6, 7, and 9 and two decimal points to write a multiplication problem with the largest possible product. Describe your strategy.

Rolling Decimals

Topic: Converting Fractions to Decimals

Object: Complete recording sheet with many different decimals.

Groups: 2 players or pair players

Materials for each group

- *Rolling Decimals* recording sheet (for each player), p. 105
- 2 number cubes (1-6)

Tip When students are ready to convert more challenging fractions, use the recording sheet for Rolling Decimals B *with two different number cubes (1–6 and 3-4-5-6-8-9).*

Directions

1. The first player rolls the two number cubes and uses the two numbers to form a proper fraction. If doubles are rolled, the player rolls again.
2. The player identifies the **proper** fraction and converts the fraction to a decimal rounded to the nearest hundredth.
3. Next the player records the rounded decimal within a correct range on her or his recording sheet. (When decimals fit within two ranges, the player has a placement choice.)
4. The second player follows these same steps. Players continue to alternate turns rolling, converting, and recording appropriately.
5. It is important for the players to realize how points are earned. Each player receives one point for every recorded decimal. (When players record more than one decimal on a range line, the decimals must be different.) The first player to record at least one decimal in each of the seven ranges receives three bonus points and the playing ends.
6. At this point each player determines her or his score. The player with the higher score wins.

.1	.2	**.22**
.21		**.34**
.28	.33 .4	**.42**
.45	.5	**.53**
.57	.6 .67	**.7**
.72		**.85**
.79		**.9**

Score: _____ 3 points (finished first)

_____ 1 point for each recorded decimal

_____ Total

Making Connections

Promote reflection and make mathematical connections by asking:

- Which range lines were more difficult to complete? Please explain.
- How would you record differently in future games?

Rolling Decimals A

.1 ______________ .22

.21 ______________ .34

.28 ______________ .42

.45 ______________ .53

.57 ______________ .7

.72 ______________ .85

.79 ______________ .9

Score: _____ 3 points (finished first)

_____ 1 point for each recorded decimal

_____
Total

Rolling Decimals B

.1 ______________ .23

.2 ______________ .35

.31 ______________ .48

.42 ______________ .58

.56 ______________ .67

.63 ______________ .77

.72 ______________ .9

Score: _____ 3 points (finished first)

_____ 1 point for each recorded decimal

_____
Total

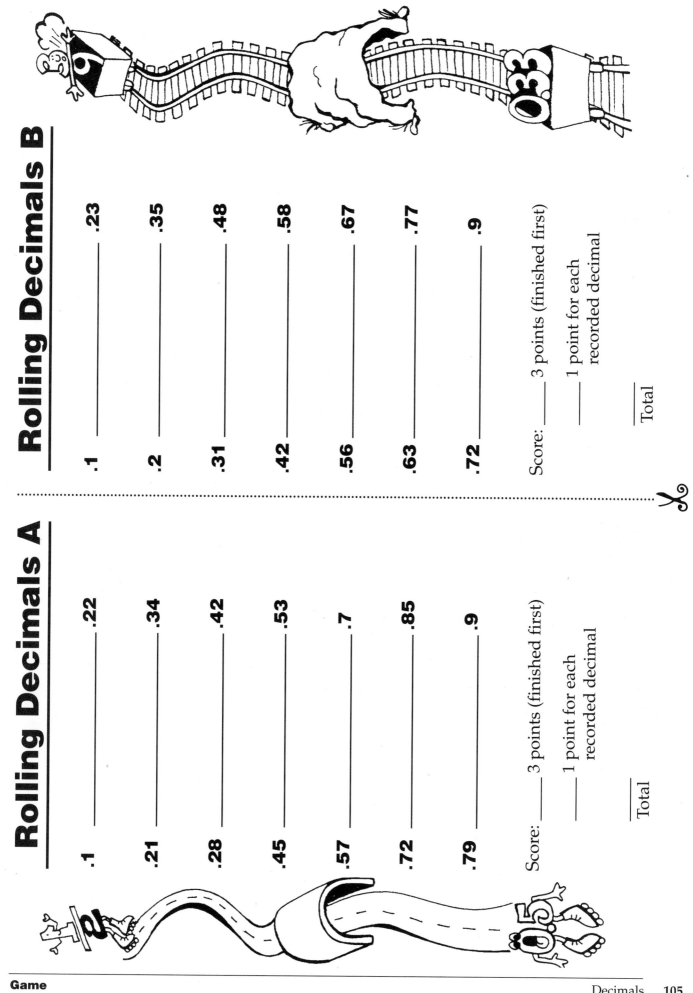

Totaling One

Topic: Mentally Adding Decimals

Object: Create the most four-in-a-rows.

Groups: 2 players or pair players

Materials for each group

- *Totaling One* gameboard, p. 107
- 25 markers for each player (2 different kinds)

Tip *Interest in this game can be extended by changing the target sum to 1.2 or an even more challenging total like .93 or 1.42.*

Directions

1. The first player covers any combination of decimals whose sum totals one with her or his markers. No more than one marker can be placed in Row 1 or in Row 2 in any one turn. The same player is required to demonstrate how the total of one was reached.
2. The second player uses her or his color markers and follows the same procedure. If a player errs in totaling one, that player removes the markers and forfeits that turn.
3. Eventually each player wants his or her placed markers arranged to display as many four-in-a-row alignments (horizontally, vertically, and diagonally) as possible.
4. Players continue to alternate turns until both players have placed all their markers or the remaining uncovered numbers will not allow a total of one with either player's remaining markers.
5. When play stops, players identify how many four-in-a-row alignments they were able to create. The winner is the player with more four-in-a-row arrangements.

Making Connections

Promote reflection and make mathematical connections by asking:

- What approach helped you locate decimal combinations that totaled one?
- What changes to the rules would you recommend?

Totaling One

	.01	.02	.03	.04	.05	.06	.07	.08	.09
.1	.11	.12	.13	.14	.15	.16	.17	.18	.19
.2	.21	.22	.23	.24	.25	.26	.27	.28	.29
.3	.31	.32	.33	.34	.35	.36	.37	.38	.39
.4	.41	.42	.43	.44	.45	.46	.47	.48	.49
.5	.51	.52	.53	.54	.55	.56	.57	.58	.59
.6	.61	.62	.63	.64	.65	.66	.67	.68	.69
.7	.71	.72	.73	.74	.75	.76	.77	.78	.79
.8	.81	.82	.83	.84	.85	.86	.87	.88	.89
.9	.91	.92	.93	.94	.95	.96	.97	.98	.99

Neighbors Count

Topic: All Decimal Operations

Object: Score the highest total.

Groups: Pair players or 2 to 4 players

Materials for each group

- *Neighbors Count* gameboard, p. 109
- 2 number cubes (1–6)
- 1 special number cube (.1–.6), p. 154
- markers (18 for each pair or player)
- scratch paper

Tip *Students enjoy scoring higher totals by also allowing points for touching cells with only a common corner.*

Directions

1. The first pair rolls the three number cubes and adds, subtracts, multiplies, and/or divides all three displayed numbers. The pair identifies and lists various possible equations. The pair locates and covers one of the resulting answers with a marker.

   Example: With 3, .4 and 5, you might compute $3 \times 5 \times .4$ and cover 6 or compute $5 - 3 + .4$ and cover 2.4.
2. The next pair rolls the number cubes and lists possible answers. Since one point is scored if the pair covers a number that **shares a side** with an already covered number, the pairs scan their lists to find answers next to the first covered cell.
3. Pairs continue to follow this process. As more numbers are covered, pairs often earn more points, since a point is scored for each adjacent covered number.

   Example: If 1.8, 2 and 2.6 are covered and you are able to cover 2.5, you would receive two points. You receive only one point if you are able to cover 2.1 or 3.
4. Pairs record a score after each turn. If a pair is unable to cover any available numbers, the pair passes for that turn.
5. The game ends after all numbers are covered or after three consecutive passes by one pair. The pair with the higher point total wins.

.1	.2	.3	.4	.5	.6
.8	1	1.2	1.2	1.4	1.5
1.5	1.6	1.8	2	2.1	2.2
2.4	2.4	2.5	2.6	3	3.2
3.5	3.6	4	4.3	4.5	4.8
5	5.4	5.5	6	7	10

Making Connections

Promote reflection and make mathematical connections by asking:

- What kinds of rolls do you prefer? Please explain.
- What worked to give you high totals on your turns?

Neighbors Count

.1	.2	.3	.4	.5	.6
.8	1	1.2	1.2	1.4	1.5
1.5	1.6	1.8	2	2.1	2.2
2.4	2.4	2.5	2.6	3	3.2
3.5	3.6	4	4.3	4.5	4.8
5	5.4	5.5	6	7	10

Date ______________ Name ______________________

Identifying Missing Decimals

Use the numbers in the box to complete each of these problems.

8.7	**56.2**
23.4	**68.8**
47.1	**74.5**

1. ____ + ____ = 83.2

2. 100 – ____ = 31.2

3. ____ + ____ = 92.2

4. 120 – ____ = 63.8

5. ____ + 20.4 + ____ = 100

6. ____ – ____ = 32.8

7. ____ – 26.3 = 20.8

8. ____ + 28.4 + ____ = 150

9. ____ – ____ = 21.7

10. ____ + ____ = 103.3

11. ____ × 5.3 = 124.02

12. 7.2 × ____ = 495.36

13. 539.52 ÷ 9.6 = ____

Use the numbers in the box to complete each of these problems.

6.9	**70.5**
34.7	**83.38**
46.1	**92.6**
59.24	**110.83**

14. ____ + ____ = 93.94

15. ____ – ____ = 18.23

16. ____ + ____ = 53

17. ____ – ____ = 11.26

18. ____ + 7.38 + ____ = 150

19. ____ + ____ = 129.48

20. ____ – ____ = 46.5

21. ____ + 19.8 + ____ = 125

22. ____ + ____ = 138.7

23. ____ – ____ = 27.45

24. ____ × ____ = 486.45

25. ____ × ____ = 3213.22

26. 3250.05 ÷ ____ = ____

27. ____ × ____ = 2446.35

28. 318.09 ÷ ____ = ____

29. 1599.67 ÷ ____ = ____

30. ____ × ____ = 638.94

31. 239.43 ÷ ____ = ____

32. 4268.86 ÷ ____ = ____

Date ____________ Name ____________

Fitting Decimals I

Use **2, 6, 4,** and **3** to complete these inequalities. Each digit may be used only once.

1. ☐☐☐.☐ < 250

2. ☐☐.☐☐ > 50

3. 3.☐ < ☐.☐ < ☐.8

4. 4.☐ > ☐.☐ < ☐.5

5. 1.☐ > 1.☐8 > 0.☐☐

6. 0.☐ < 0.5☐ < 0.☐☐

Use **3, 5, 7,** and **4** to complete these inequalities. Each digit may be used only once.

7. ☐.☐ + 0.☐☐ < 5

8. 5.☐☐ − 0.☐☐ > 5

9. 8.☐ + 0.☐☐ < 9.☐

10. ☐.☐ − 0.☐ > 6.☐

11. 7.☐ + ☐.☐ > 14.☐

12. 0.☐ − 0.☐☐ > .2☐

Use **3, 9, 7,** and **6** to create the largest possible answer. Each digit may be used only once.

13. ☐☐.7 + ☐.☐ = ____

14. 4☐.☐ + ☐.☐5 = ____

15. ☐.6 + .7☐ + ☐.☐9 = ____

16. ☐☐.6 − ☐☐.9 = ____

17. ☐.2☐ − 0.☐☐ = ____

18. ☐.4☐ − 3.3☐ + 0.☐5 = ____

Use **2, 4, 7,** and **6** to create the smallest possible answer. Each digit may be used only once.

19. ☐☐.5 − ☐.☐ = ____

20. 3☐.☐ + ☐☐.8 = ____

21. 4.☐☐ − ☐.7☐ = ____

22. ☐☐.3 − 5.☐☐ = ____

23. ☐.5 + 0.☐☐ − ☐.9 = ____

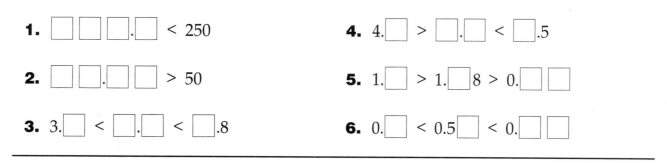

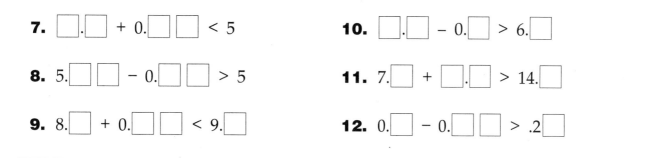

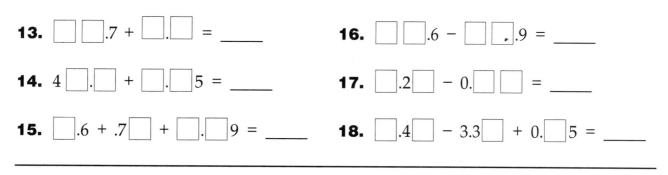

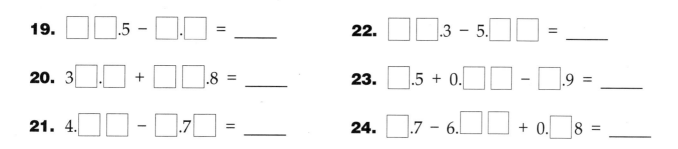

24. ☐.7 − 6.☐☐ + 0.☐8 = ____

Date ____________ Name ____________________

Fitting Decimals II

Use **2, 4, 7,** and **9** to complete these inequalities. Each digit may be used only once.

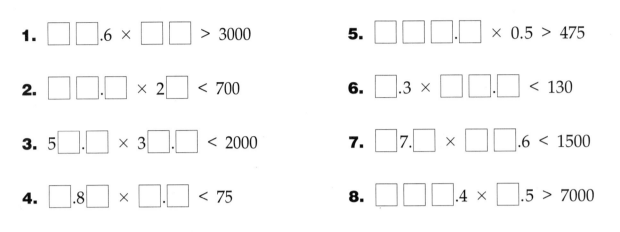

1. □□.6 × □□ > 3000

2. □□.□ × 2□ < 700

3. 5□.□ × 3□.□ < 2000

4. □.8□ × □.□ < 75

5. □□□.□ × 0.5 > 475

6. □.3 × □□.□ < 130

7. □7.□ × □□.6 < 1500

8. □□□.4 × □.5 > 7000

Use **4, 5, 7,** and **8** to create the largest possible answer. Each digit may be used only once.

9. □.□ × □.□ = _____

10. □□.7 × □.□ = _____

11. □□.□ × 0.□ = _____

12. 0.□□ × □.□ = _____

Use **3, 6, 8,** and **5** to create the smallest possible answer. Each digit may be used only once.

13. □.□ × □.□ = _____

14. □□.□ × 0.□ = _____

15. □□.□ × 6.□ = _____

16. 4.□□ × □.□ = _____

Use **3, 2, 8,** and **6** to complete each of these problems. Some of the digits have been filled in the boxes for you. Each digit may be used only once.

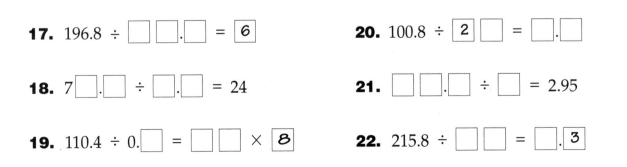

17. 196.8 ÷ □□.□ = [6]

18. 7□.□ ÷ □.□ = 24

19. 110.4 ÷ 0.□ = □□ × [8]

20. 100.8 ÷ [2]□ = □.□

21. □□.□ ÷ □ = 2.95

22. 215.8 ÷ □□ = □.[3]

Percents

Assumptions Percent concepts and calculations have previously been taught and reviewed, emphasizing understanding and building number sense. Concrete objects and visual models, such as base 10 materials and grids, have been used extensively.

Section Overview and Suggestions

Sponges

Placing Percents pp. 114–115

Target 100% pp. 116–117

Ordering Percents, Decimals, and Fractions p. 118

These whole-class or small-group warm-ups are repeatable. They build percent and number sense as students enhance their abilities to convert and compute percents mentally and with paper and pencil. Frequent and repeated use of these Sponges will ensure greater success with all the games and independent activities in this section.

Skill Checks

Partial Possibilities 13–16 pp. 119–121

The Skill Checks provide a way for parents, students, and you to see students' improvement with percent conversion and computation. Copies can be cut in half so that each check may be used at a different time. Remember to have all students respond to the STOP, number sense task, before solving the ten problems.

Games

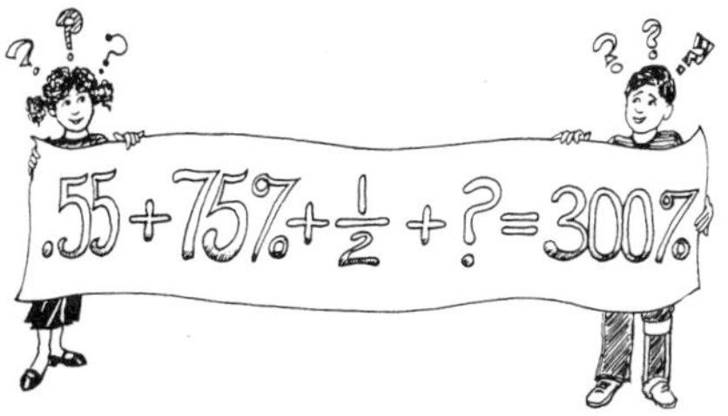

Target 300% pp. 122–123

Figuring Percents pp. 124–126

Take the Discount pp. 127–128

These open-ended and repeatable Games actively involve students in converting and comparing percents, fractions, and decimals, as students enhance their strategic thinking abilities. To ensure more active and successful participation with *Target 300%*, provide adequate practice with the *Target 100%* sponge.

Independent Activities

Finding Tic-Tac-Toes p. 129

Figuring Percents Practice pp. 130–131

Fitting Percents p. 132

These engaging activities require students to independently practice converting fractions and decimals to percents and computing percents. Use of *Figuring Percents Practice* should ensure greater success with the *Figuring Percents* game. These challenging activities involve mental computation, as students improve their number sense.

Placing Percents

Tip *As students gain competence and confidence, include the "7" and "0" Digit Squares.*

Topic: Comparing Percents, Decimals, and Fractions

Object: Create qualifying expressions involving percents, decimals, fractions, and inequalities.

Groups: Whole class or small group

Materials

- *Placing Percents* activity form, p. 115
- set of Digit Squares (0 & 7 removed), p. 151
- scratch paper for each participant

Directions

1. After the leader selects and displays the first sentence from the *Placing Percents* activity form, students copy the displayed format.
2. The leader draws and displays two Digit Squares.
3. Each student appropriately records each of the displayed digits in two of the four spaces.

   Example: 1 and 4 are drawn. One student records . __ 4 > 1 __ %, while another player records . 4__ > 1__ %.
4. The leader draws and displays two additional Digit Squares which students attempt to appropriately record in their sentences. The leader has students share different solutions.
5. The used Digit Squares are returned and the process is repeated using the same displayed format. The leader has students share their varied solutions.
6. Next the leader displays a new format, such as the second or third sentence, which students record. A similar process of drawing Digit Squares is followed.
7. After students practice comparing decimals and fractions to percents, the fifth sentence is displayed and copied by the participants. Since six digits are required for each of the remaining formats, the leader draws two Digit Squares three times for each sentence. A similar process of displaying, ordering, recording and sharing of results continues.

Making Connections

Promote reflection and make mathematical connections by asking:

- Which expressions were easier to create? Why?
- What approach helped you successfully place digits?

Placing Percents

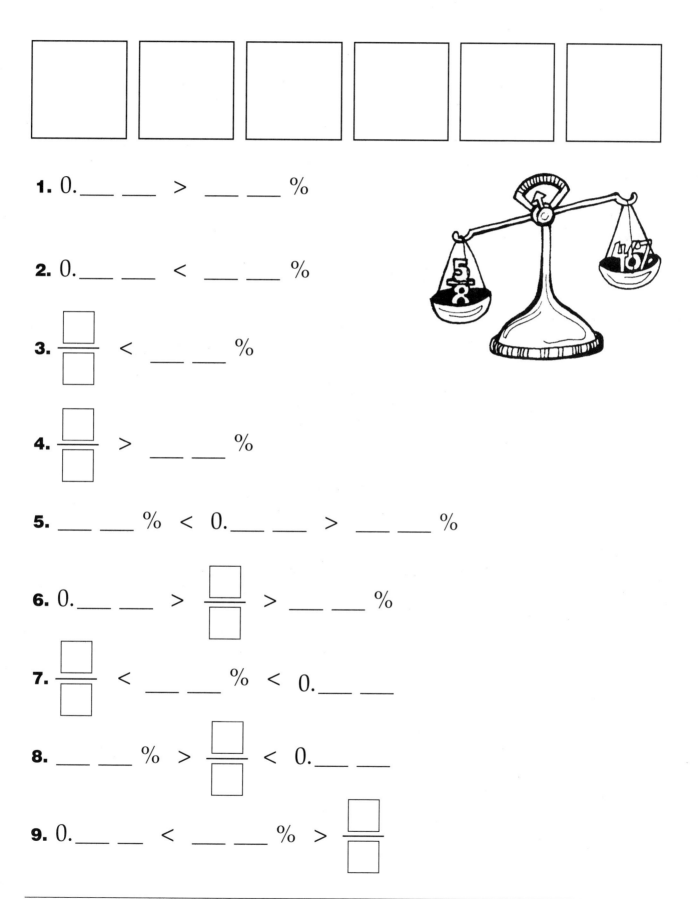

1. 0.___ ___ > ___ ___ %

2. 0.___ ___ < ___ ___ %

3. $\frac{\square}{\square}$ < ___ ___ %

4. $\frac{\square}{\square}$ > ___ ___ %

5. ___ ___ % < 0.___ ___ > ___ ___ %

6. 0.___ ___ > $\frac{\square}{\square}$ > ___ ___ %

7. $\frac{\square}{\square}$ < ___ ___ % < 0.___ ___

8. ___ ___ % > $\frac{\square}{\square}$ < 0.___ ___

9. 0.___ ___ < ___ ___ % > $\frac{\square}{\square}$

Target 100%

Topic: Mental Addition of Fractions and Percents

Object: Total exactly 100% or close to it.

Groups: Whole class or small group

Materials

- transparency of *Target 100%* activity form, p. 117
- 16 markers (8 of two contrasting colors)

Tip This sponge can be easily adapted into a game for pair players or two players. The Target 300% *game, pp. 122–123, provides a more challenging version.*

Directions

1. Divide the class or group into two teams. Each team uses its own colored markers to alternately cover cells.
2. One member from the first team places a marker on a cell and announces the covered percent (converting fractions to percents when necessary).
3. The second team places a different colored marker on any uncovered cell and announces that percent.
4. A new member from the first team selects and covers another cell, adding this percent to the team's previously covered percent, and states the total.

   Example:
   A team having covered $\frac{1}{10}$ and 15% announces, "10 percent and 15 percent equals twenty-five percent."

Team A	Team B
10%	20%
15%	25%

5. Teams alternate turns by placing a marker on an uncovered cell and stating the team's accumulated total.
6. The team that reaches and states "100 percent" or the percent closer to 100% wins.

15%	$\frac{1}{5}$	5%	10%
$\frac{1}{10}$	$\frac{1}{20}$	20%	$\frac{1}{4}$
$\frac{2}{10}$	35%	$\frac{3}{10}$	$\frac{3}{20}$
5%	$\frac{3}{12}$	25%	$\frac{2}{20}$

Making Connections

Promote reflection and make mathematical connections by asking:

- How would you place markers differently in future rounds?
- What strategy helped teams reach 100% or close to it?

Target 100%

15%	$\frac{1}{5}$	5%	10%
$\frac{1}{10}$	$\frac{1}{20}$	20%	$\frac{1}{4}$
$\frac{2}{10}$	35%	$\frac{3}{10}$	$\frac{3}{20}$
5%	$\frac{3}{12}$	25%	$\frac{2}{20}$

Ordering Percents, Decimals, and Fractions

Topic: Comparing Values of Percents, Decimals, and Fractions

Object: Order fractions, decimals, and percents from least to greatest.

Groups: Whole class or small group

Tip *Increase difficulty by requiring the ordering of more than eight of the displayed fractions, decimals, and percents.*

Materials

- transparent set of Digit Squares (0, 7 and 9 removed), p. 151
- chalkboard or overhead projector

Directions

1. The leader draws and displays three Digit Squares. Students select any two of the three digits to express, record, and display three different percents.
2. Students then use any two of the three digits to express, record, and display three decimals that are not equivalent to the displayed percents.
3. Next students use the same displayed digits to express fractions. The possible fractions are listed and displayed.

   Example: 5, 3 and 8 are displayed.

   If 35%, 38% and 83% are chosen for percents, .58, .53 and .85 would be displayed for decimals. The possible fractions are $\frac{3}{5}$, $\frac{3}{8}$, $\frac{5}{8}$, $\frac{5}{3}$, $\frac{8}{3}$, and $\frac{8}{5}$.
4. The group is divided into two teams. The leader displays eight horizontal lines and states that eight of the displayed percents, decimals, and fractions need to be recorded in order from least to greatest.
5. The first team selects one of the displayed amounts and chooses its placement on one of the eight lines. That amount is crossed out and recorded.
6. The second team places one of the remaining displayed percents, decimals, or fractions on another line. If the least to greatest order is not followed, the first team automatically receives a point and the activity begins again.
7. The teams take turns, picking numbers from the remaining choices and placing them in order.

   ___ ___ 38% ___ ___ .53 ___ $\frac{8}{3}$

   least ... greatest
8. If the teams are able to correctly order eight different displayed amounts, each team receives two points. If a team is able to prove that none of the remaining amounts may be placed on an unfilled line, that team receives one point and the round ends.

Making Connections

Promote reflection and make mathematical connections by asking:

- What helped you order the fractions with the percents and decimals?
- What strategy works well to ensure all eight lines will be used?

Date ____________ Name ____________________

Partial Possibilities 13

STOP Don't start yet. Star a problem that may have an answer with the smallest percentage.

1. $\frac{1}{5} = \_\_\_\_\_ \%$

2. $80\% = \frac{\square}{\square}$ lowest terms

3. 20% of 15 = _____

4. 35% of 22 = _____

5. $\frac{3}{4} + \_\_\_\_\_ \% = 80\%$

6. $\frac{1}{4} + 0.35 + \_\_\_\_\_ \% = 100\%$

7. Order from least to greatest.
$\frac{1}{3}$, 0.40, 37% ______________

8. $30 less 10% = $ _____

Place 6, 7, and 3 in each equation.

9. $\frac{\square}{\square} < 5\% \square$

10. $\square\square \% > \frac{\square}{4}$

Go On Which is greater, 20% of 85 or 40% of 50? Prove it.

Date ____________ Name ____________________

Partial Possibilities 14

STOP Don't start yet. Star a problem that may have an answer less than 50%.

1. $\frac{3}{4} = \_\_\_\_\_ \%$

2. $28\% = \frac{\square}{\square}$ lowest terms

3. 25% of 80 = _____

4. 15% of 30 = _____

5. $\frac{2}{5} + \_\_\_\_\_ \% = 75\%$

6. $\frac{1}{5} + 0.52 + \_\_\_\_\_ \% = 100\%$

7. Order from least to greatest.
$\frac{7}{8}$, 72%, 0.59 ______________

8. $48 less 50% = $ _____

Place 2, 5, and 9 in each equation.

9. $\frac{\square}{\square} > 4\square\%$

10. $\square\square \% < \frac{3}{\square}$

Go On Estimate 11% of 48 and justify your estimate.

Date ______________ Name ______________________

Partial Possibilities 15

STOP Don't start yet. Star a problem that may have a whole-number answer.

1. $\frac{7}{10}$ = _____ %

2. 20% = $\frac{\square}{\square}$ lowest terms

3. 40% of 30 = _____

4. 24% of 40 = _____

5. $\frac{1}{2}$ + _____ % = 80%

6. $\frac{3}{4}$ + 0.10 + _____ % = 100%

7. Order from least to greatest.
$\frac{5}{9}$, 32%, 0.62 ______________

8. $70 less 20% = $ _____

Place 3, 5, and 8 in each equation.

9. $\frac{\square}{\square} < \square$ 2%

10. $\square\square$% > $\frac{\square}{5}$

Go On Which is greater, 30% of 120 or 60% of 50? Prove it.

Date ______________ Name ______________________

Partial Possibilities 16

STOP Don't start yet. Star a problem that may have an answer close to one-half.

1. $\frac{1}{4}$ = _____ %

2. 55% = $\frac{\square}{\square}$ lowest terms

3. 60% of 40 = _____

4. 26% of 30 = _____

5. $\frac{1}{5}$ + _____ % = 75%

6. $\frac{2}{5}$ + 0.43 + _____ % = 100%

7. Order from least to greatest.
53%, $\frac{3}{7}$, 0.35 ______________

8. $10 less 15% = $ _____

Place 2, 6, and 8 in each equation.

9. $\frac{\square}{\square} < \square$ 5%

10. $\square\square$% > $\frac{\square}{3}$

Go On Estimate 24% of 63 and justify your estimate.

Date ____________________ Name ________________________________

Partial Possibilities 17

STOP Don't start yet. Star a problem that may have an answer less than 20%.

1. $\frac{1}{2}$ = _____ %

2. 40% = $\boxed{\frac{\ }{\ }}$ lowest terms

3. 75% of 60 = _____

4. 38% of 35 = _____

5. $\frac{3}{5}$ + _____ % = 75%

6. $\frac{4}{5}$ + 0.15 + _____ % = 100%

7. Order from least to greatest.
24%, 0.33, $\frac{1}{6}$ ______________

8. $40 less 25% = $ _____

Place 3, 4, and 7 in each equation.

9. $\frac{\square}{\square} < \square 8\%$

10. $\square\square\% < \frac{\square}{5}$

Go On Which is greater, 40% of 90 or 80% of 40? Prove it.

Date ____________________ Name ________________________________

Partial Possibilities 18

STOP Don't start yet. Star a problem that may have a decimal answer.

1. $\frac{2}{5}$ = _____ %

2. 75% = $\boxed{\frac{\ }{\ }}$ lowest terms

3. 30% of 50 = _____

4. 18% of 60 = _____

5. $\frac{1}{4}$ + _____ % = 80%

6. $\frac{3}{5}$ + 0.17 + _____ % = 100%

7. Order from least to greatest.
$\frac{8}{9}$, 0.58, 71% ______________

8. $55 less 20% = $ _____

Place 4, 5, and 8 in each equation.

9. $\frac{\square}{\square} < 6\square\%$

10. $\square\square\% < \frac{3}{\square}$

Go On Estimate 34% of 59 and justify your estimate.

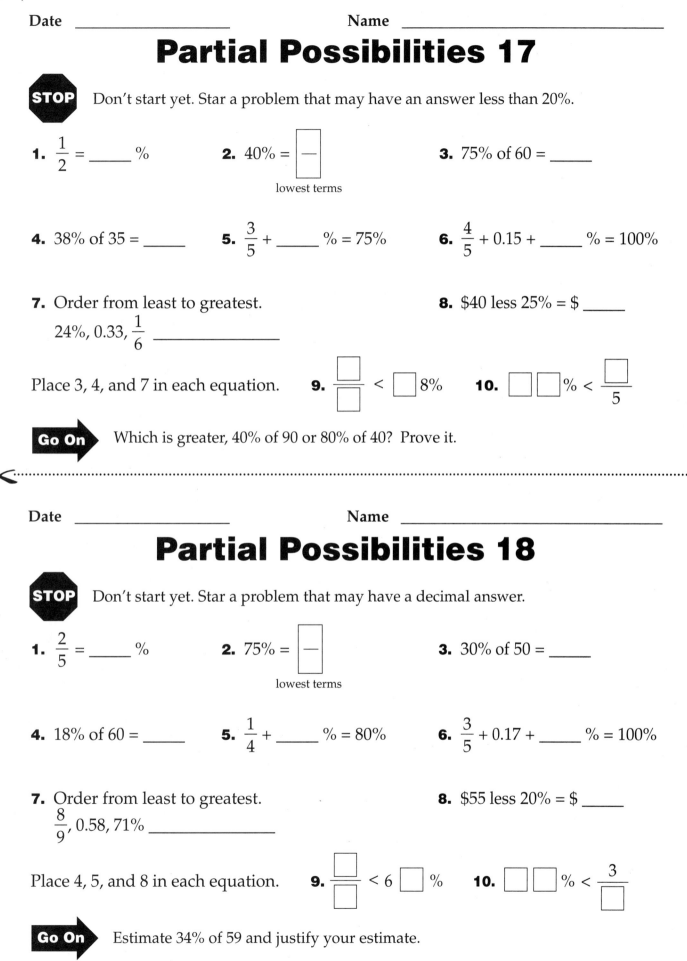

Target 300%

Topic: Mental Addition of Fractions, Decimals, and Percents

Object: Total exactly 300% or close to it.

Groups: 2 pair players or two players

Materials for each group

- *Target 300%* gameboard, p. 123
- 16 markers (8 each of two contrasting colors)

Tip *Encourage playing additional rounds so players see patterns and discover winning strategies.*

Directions

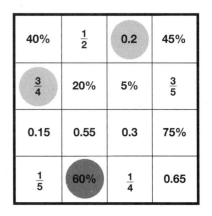

40%	$\frac{1}{2}$	0.2	45%
$\frac{3}{4}$	20%	5%	$\frac{3}{5}$
0.15	0.55	0.3	75%
$\frac{1}{5}$	60%	$\frac{1}{4}$	0.65

1. The first pair places a marker on a cell and announces the covered percent (converting fractions and decimals to percents when necessary).
2. The second pair places a different colored marker on any uncovered cell and announces that percent.
3. The first pair selects and covers another cell. This amount is added to the pair's previously covered amount. The pair then states the total percent.
   Example: A pair, having covered $\frac{3}{4}$ and then .2, announces "75 percent and 20 percent equals 95 percent."
4. Pairs alternate turns by placing markers on uncovered cells and stating their accumulated total.
5. The pair who reaches and states "300 percent" or the percent closer to 300% wins.

Making Connections

Promote reflection and make mathematical connections by asking:

- When and why did you find it necessary to block your opponent?
- What strategy helped pairs reach 300% or close to it?

Target 300%

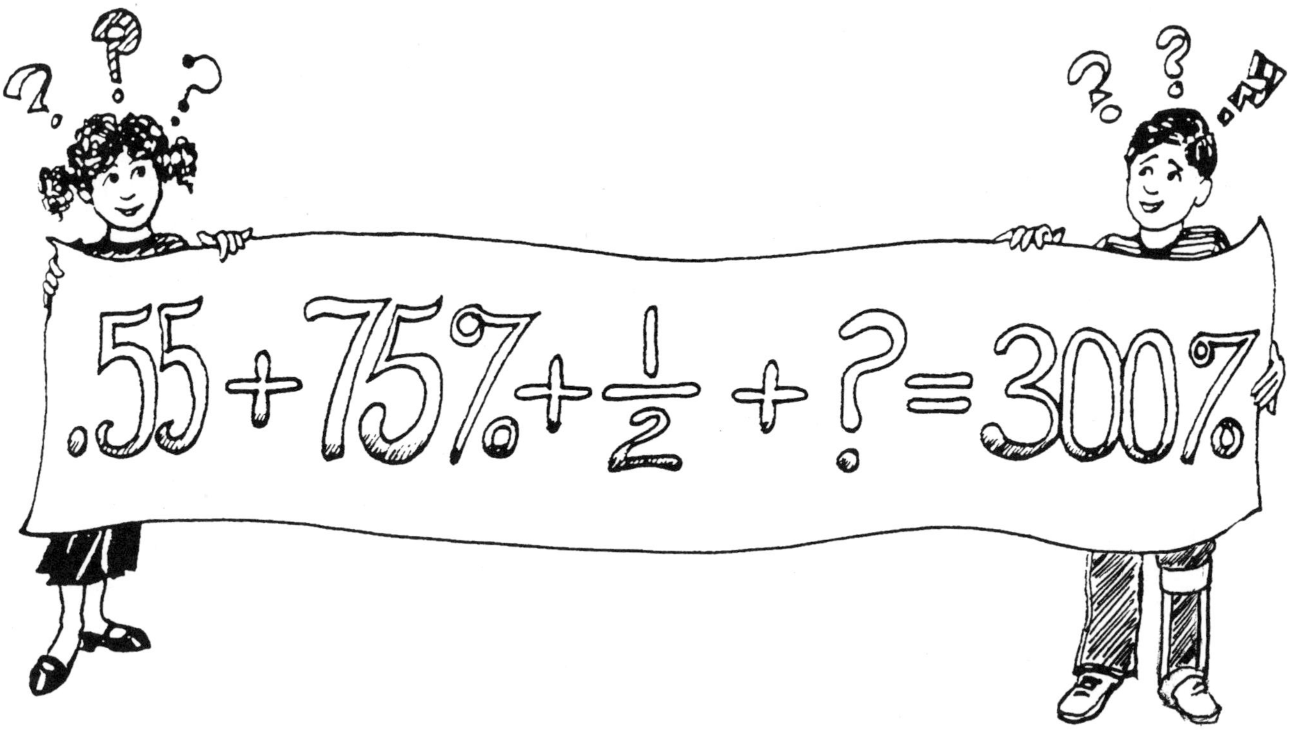

40%	$\frac{1}{2}$	0.2	45%
$\frac{3}{4}$	20%	5%	$\frac{3}{5}$
0.15	0.55	0.3	75%
$\frac{1}{5}$	60%	$\frac{1}{4}$	0.65

Figuring Percents

Topic: Mental Computation of Percents

Object: Cover four numbers in a row.

Groups: 2 players or pair players

Materials for each group

- *Figuring Percents* gameboard, p. 125
- markers (different kind for each player)
- calculator

Tip *When students become confident with this game, consider using the more challenging choices on the* Figuring Percents B *gameboard, p. 126.*

Directions

6	3	2.5	8	5
4	1.2	4.8	3	7.5
2	3.6	1	6	2
12	6	3	3.75	4
3	4.5	9	1.5	2.4

Choices				
10	12	15	20	30
10%	20%	25%	30%	40%

1. The first player selects and announces a whole number and a percent (one from each row in the "Choices" box). After the choices are announced, the player is allowed to use a calculator or pencil and paper to find the product. The player verbally states the problem with the resulting value and covers that amount on the gameboard with his or her colored marker. If the value appears more than once on the gameboard, only one square may be covered each turn.

   Example: If a player selects 30% and 15, the player states, "30% of 15 equals four and five tenths."
2. The second player selects and announces one whole number and one percent from the "Choices" box and finds and checks the product with a calculator or paper and pencil. If the resulting amount is not covered, it is covered by that player's marker. If all the squares containing that amount are covered, the player forfeits that turn.
3. Players continue alternating turns, announcing choices, finding values of percents, and covering the resulting amount on the gameboard.
4. The first player to have four markers in a row horizontally, vertically, or diagonally wins.

Making Connections

Promote reflection and make mathematical connections by asking:

- What approach helped you mentally compute these percents?

Figuring Percents

Choices

10	12	15	20	30
10%	20%	25%	30%	40%

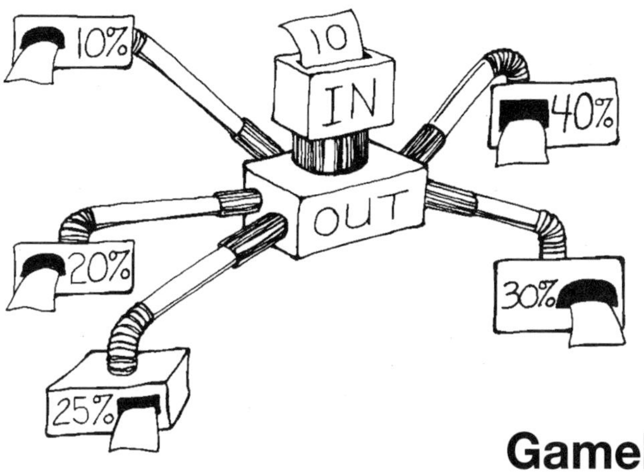

Gameboard A

6	3	2.5	8	5
4	1.2	4.8	3	7.5
2	3.6	1	6	2
12	6	3	3.75	4
3	4.5	9	1.5	2.4

Figuring Percents

Choices				
15	24	31	35	66
15%	20%	30%	35%	49%

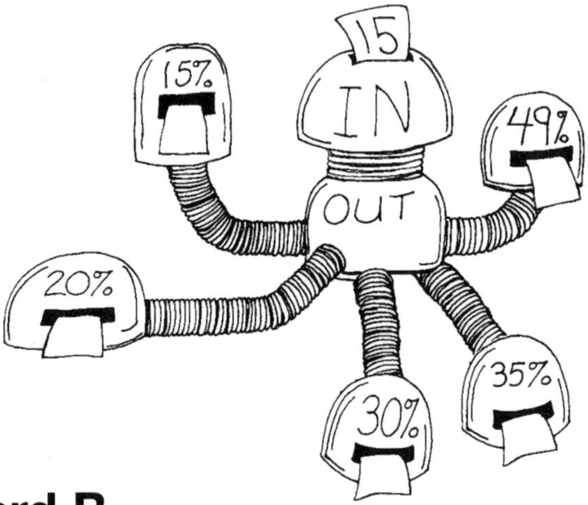

Gameboard B

10.85	9.3	5.25	4.8	7.35
7	19.8	2.25	11.76	8.4
5.25	17.15	3.6	6.2	3
7.2	4.65	15.19	10.5	13.2
32.34	23.1	4.5	9.9	12.25

Take the Discount

Topic: Computing Discounted Prices

Object: Cover four numbers in a row.

Groups: 2 players or pair players

Materials for each group

- *Take the Discount* gameboard, p. 128
- markers (different kind for each player)
- paper clip and pencil for spinner
- scratch paper

Tip An interesting variation is to have both players use the same spin but indicate different full prices to determine placement of their markers.

Directions

1. The first player spins the spinner on the gameboard, then selects one of the displayed dollar amounts. The player reduces the indicated price by the spun amount to determine where to place her or his colored marker.

   Example: The player spins "less 40%" and indicates \$30. The player states, "40% of \$30 is \$12, so the discounted price is \$18."
2. The second player spins, indicates an initial price, and computes to determine the discount and discounted price, and where to place her or his marker. Players are allowed to share a cell and use shared cells to produce four-in-a-row arrangements.
3. Players continue to alternate turns following this process of spinning, identifying full prices, figuring discounts and discounted prices, and covering the results.
4. The first player to get four in a row horizontally, vertically, or diagonally wins.

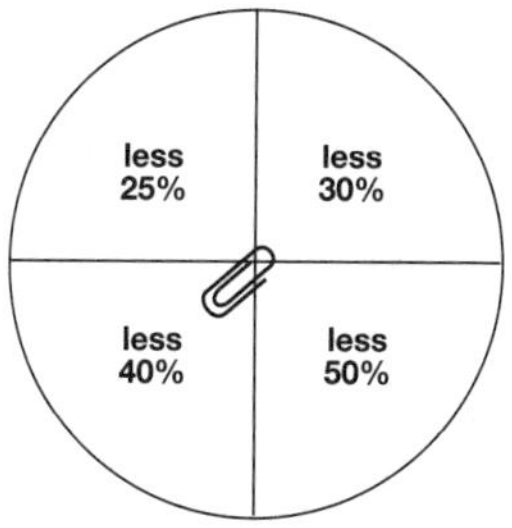

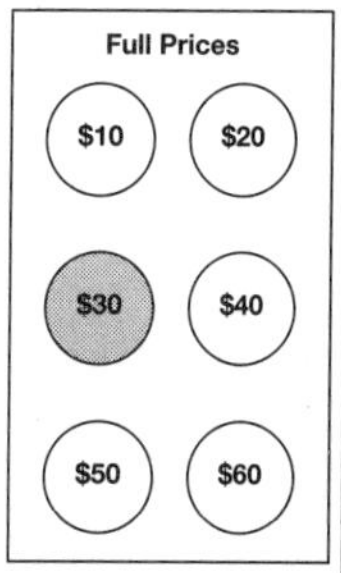

\$5.00	\$22.50	\$18.00	\$35.00	\$10.00
\$30.00	\$12.00	\$7.00	\$15.00	\$36.00
\$45.00	\$24.00	\$30.00	\$6.00	\$20.00
\$14.00	\$7.50	\$37.50	\$42.00	\$30.00
\$21.00	\$30.00	\$15.00	\$25.00	\$28.00

Making Connections

Promote reflection and make mathematical connections by asking:

- How did sharing of cells change your playing strategy?
- Which discounted prices were easy to compute mentally? Please explain.

Take the Discount

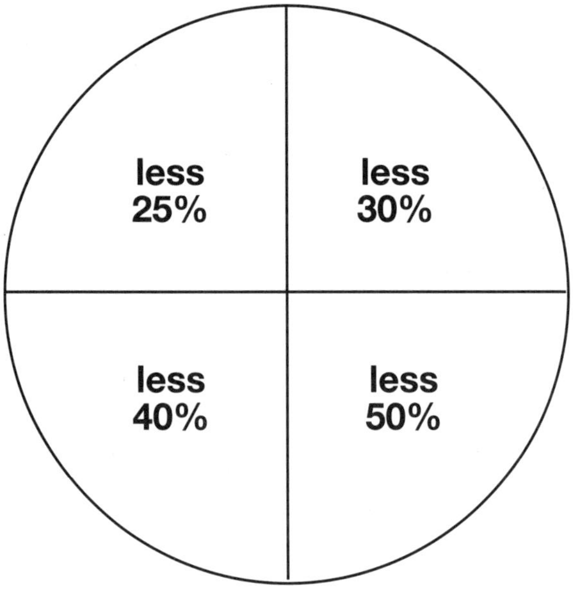

Full Prices

$10	$20
$30	$40
$50	$60

$5.00	$22.50	$18.00	$35.00	$10.00
$30.00	$12.00	$7.00	$15.00	$36.00
$45.00	$24.00	$30.00	$6.00	$20.00
$14.00	$7.50	$37.50	$42.00	$30.00
$21.00	$30.00	$15.00	$25.00	$28.00

Date ____________________ Name ________________________________

Finding Tic-Tac-Toes

For each problem, complete the equations on the left. Find the Tic-Tac-Toe in the grids on the right by checking to see which row, column, or diagonal contains three numbers which match answers from the equations.

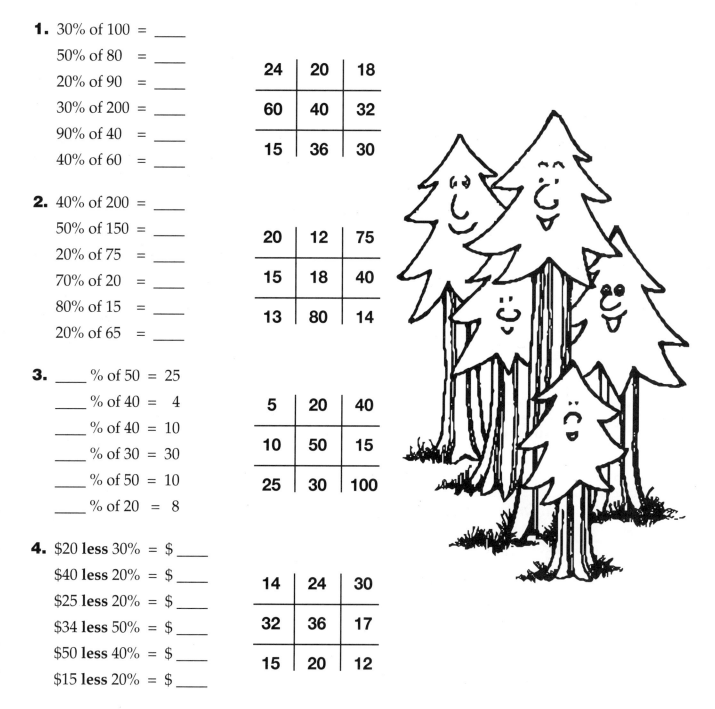

1. 30% of 100 = ____
50% of 80 = ____
20% of 90 = ____
30% of 200 = ____
90% of 40 = ____
40% of 60 = ____

24	20	18
60	40	32
15	36	30

2. 40% of 200 = ____
50% of 150 = ____
20% of 75 = ____
70% of 20 = ____
80% of 15 = ____
20% of 65 = ____

20	12	75
15	18	40
13	80	14

3. ____ % of 50 = 25
____ % of 40 = 4
____ % of 40 = 10
____ % of 30 = 30
____ % of 50 = 10
____ % of 20 = 8

5	20	40
10	50	15
25	30	100

4. $20 **less** 30% = $____
$40 **less** 20% = $____
$25 **less** 20% = $____
$34 **less** 50% = $____
$50 **less** 40% = $____
$15 **less** 20% = $____

14	24	30
32	36	17
15	20	12

TRIVIA The middle number in the Tic-Tac-Toe row for #4 equals the number of trees saved by 1 ton of recycled paper.

Challenge Create a similar Tic-Tac-Toe problem for your classmates to solve. Be sure your puzzle has only one Tic-Tac-Toe row.

Date ______________ Name ______________

Figuring Percents Practice I

Create multiplication problems and answers which match the descriptions below by using a percent from one box and a whole number from the other box.

25%	50%	75%

40	60	80

1. ______________ odd product

2. ______________ product greater than 50

3. ______________ product between 40 and 50

4. ______________ product less than 30

10%	30%	60%

40	50	70

5. ______________ smallest even product

6. ______________ product greater than 30

7. ______________ product less than 10

8. ______________ product that is a multiple of 10

15%	40%	60%

20	75	80

9. ______________ largest possible product

10. ______________ product between 25 and 35

11. ______________ product less than 10

12. ______________ odd product greater than3 0

Date ____________ Name ____________

Figuring Percents Practice II

Create multiplication problems and answers which match the descriptions below by using a percent from one box and a whole number from the other box.

35%	**55%**	**85%**

80	**400**	**500**	**700**

1. ____________
largest possible even product

2. ____________
smallest possible odd product

3. ____________
product between 200 and 300

4. ____________
product greater than 450

40%	**65%**	**90%**

600	**750**	**800**	**1000**

5. ____________
largest possible odd product

6. ____________
product between 600 and 700

7. ____________
multiple of 100

8. ____________
product greater than 750

30%	**60%**	**70%**

360	**480**	**720**	**960**

9. ____________
product between 500 and 600

10. ____________
product less than 250

11. ____________
product between 250 and 350

12. ____________
product between 400 and 500

Date ______________ Name ______________________

Fitting Percents

Use the given digits only once in each equation or inequality.

Use **1, 3, 5, 6,** and **8** to complete these inequalities.

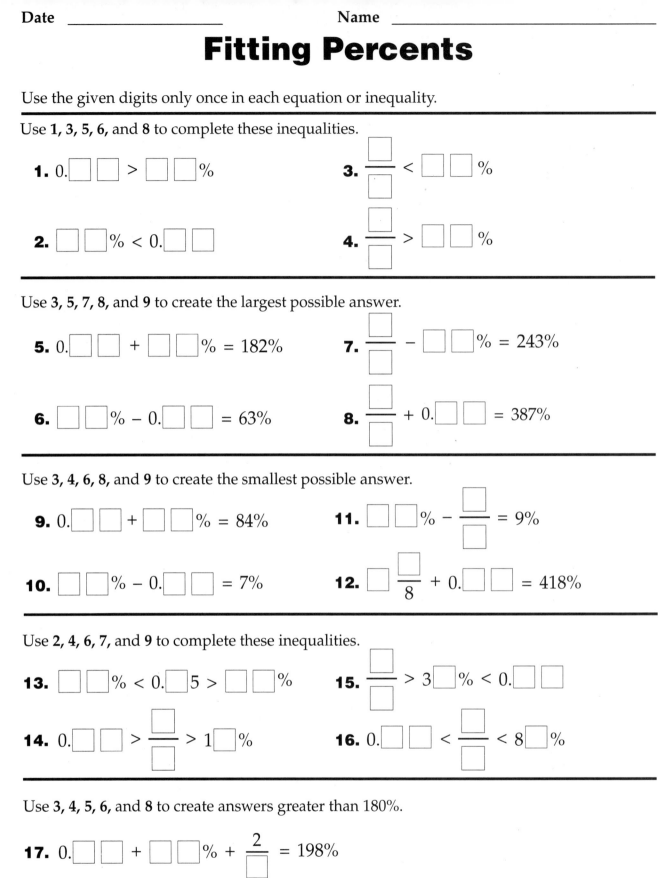

1. $0.\square\square > \square\square\%$

2. $\square\square\% < 0.\square\square$

3. $\frac{\square}{\square} < \square\square\%$

4. $\frac{\square}{\square} > \square\square\%$

Use **3, 5, 7, 8,** and **9** to create the largest possible answer.

5. $0.\square\square + \square\square\% = 182\%$

6. $\square\square\% - 0.\square\square = 63\%$

7. $\frac{\square}{\square} - \square\square\% = 243\%$

8. $\frac{\square}{\square} + 0.\square\square = 387\%$

Use **3, 4, 6, 8,** and **9** to create the smallest possible answer.

9. $0.\square\square + \square\square\% = 84\%$

10. $\square\square\% - 0.\square\square = 7\%$

11. $\square\square\% - \frac{\square}{\square} = 9\%$

12. $\square\frac{\square}{8} + 0.\square\square = 418\%$

Use **2, 4, 6, 7,** and **9** to complete these inequalities.

13. $\square\square\% < 0.\square 5 > \square\square\%$

14. $0.\square\square > \frac{\square}{\square} > 1\square\%$

15. $\frac{\square}{\square} > 3\square\% < 0.\square\square$

16. $0.\square\square < \frac{\square}{\square} < 8\square\%$

Use **3, 4, 5, 6,** and **8** to create answers greater than 180%.

17. $0.\square\square + \square\square\% + \frac{2}{\square} = 198\%$

18. $0.\square 9 - \square\square\% + \frac{\square}{\square} = 216\%$

Algebra Readiness

Assumptions Number theory topics and simple algebraic concepts and equations have been introduced and reviewed, emphasizing understanding and enhancing number sense. Counters and a variety of visual models, such as factor trees and number lines, have been used extensively.

Section Overview and Suggestions

Sponges

Which Doesn't Belong? p.134

Common Factors Match pp. 135–136

Identifying Values p. 137

What's My Number? p.138

These whole-class or small-group warm-ups are repeatable. They reinforce number concepts and promote mathematical reasoning as students improve their abilities to represent numerical situations symbolically and solve simple algebraic equations. Repeat use of these Sponges will ensure greater success with the Games and Independent Activities in this section.

Skill Checks

Number Explorations 1–6 pp. 139–141

The Skill Checks provide a way for parents, students, and you to see students' improvement with number theory topics and introductory algebra concepts. Copies can be cut in half so checks may be used at different times. Remember to have all students respond to the STOP, number sense task, before solving the ten problems.

Games

Greatest Common Factor Bingo pp. 142–143

Sixty p.144

Seek and Cover pp. 145–146

These open-ended and repeatable Games actively involve students in enhancing skills for algebra success. *Greatest Common Factor Bingo* and *Sixty* reinforce factors, multiples and unique numbers. Use of the *What's My Number?* sponge should ensure student success when solving for unknowns in *Seek and Cover.* With frequent use of these games at home or in family math sessions, students will gain confidence in their abilities to do more abstract levels of mathematics.

Independent Activities

Number Riddle Arrangements p. 147

Discover Trivia by Identifying Unknowns p. 148

These activities engage students in independent practice of applying number theory concepts and solving simple algebraic equations. *Number Riddle Arrangements* encourages students to create similar riddles for classmates to solve, providing additional high-interest practice of more abstract, important math skills.

Which Doesn't Belong?

Topic: Number Theory Concepts

Object: Identify a common characteristic shared by three of four numbers.

Groups: Whole class or small group

Materials

- prepared starter problems
- chalkboard or overhead projector

Tip *Students appreciate repeat rounds and multiple-day experiences using classmate-authored puzzles.*

Directions

16	49
56	36

1. The leader displays four numbers using the illustrated format and asks, "Which one doesn't belong?"
2. As students study the example, they should look for common characteristics that only three of the numbers share.
3. The leader calls on a student to identify a number that does not belong. The leader then asks the entire group to determine characteristics the identified number does not share with the other three numbers.
4. The first student describes a characteristic shared by only three of the numbers. Students add their reasons why the number doesn't belong.
   Example: A student identifies 49 as not belonging. Students may point out that the other numbers have a 6 in the one's place or that they are even numbers. The identifying student might explain that 49 doesn't belong because the other three are multiples of 4 and 49 is not.
5. The leader asks which different numbers might not belong and follows a similar process, continuing until no other number is identified as not belonging. (In the above example, 56 might not belong because the others are square numbers or 36 might not belong because the sums of the digits of the other three numbers equal prime numbers and the sum of 3 and 6 equals a composite number.)
   Additional sample problems:

15	4
36	12

Factors of 60

56	84
34	21

Multiples of 7

19	43
57	31

Primes

56	81
144	100

Square Numbers

243	27
160	39

Divisible by 3

6. After two or three samples, students pair up or work independently to create similar problems for classmates to solve. It is important that the creators identify at least one common characteristic shared only by three of the four numbers.

Making Connections

Promote reflection and make mathematical connections by asking:

- How does this warm-up promote mathematical thinking?
- What approaches work for creating similar puzzles that promote lots of possibilities?

Common Factors Match

Topic: Common Factors

Object: Place numbers so every number shares a common factor with every adjacent number.

Groups: Whole class or small group

Materials

- transparency of *Common Factors Match* recording grid, p. 136
- overhead pens

Tips After a round is tried, students can partner up to attempt the activity as pairs. This warm-up activity can be simplified by using a 3×3 grid and fewer numbers.

Directions

1. Class is divided into two or more groups. The goal is for the groups to collaborate and place all sixteen displayed numbers in one recording grid so that every number shares a common factor with each adjacent cell, horizontally, vertically, and diagonally.
2. The groups take turns selecting and placing numbers. One volunteer from the first group selects one number to place in the recording grid.
   Example: "We choose 6. Please record it in the cell located in the second row and third column." (Six is crossed out after it is recorded in the identified cell.)
3. One volunteer from the second group identifies another number and where to record it in the grid. This second number must share a common factor with the previously placed number and adjoin that number.
4. A volunteer from the next group selects a number and places it next to any previously recorded number, being sure that it shares at least one common factor with each number in an adjacent cell.
5. The groups continue to take turns selecting and cautiously placing numbers.
6. If a number is recorded that does not share a common factor with each adjacent cell, the number is removed and replaced as a number choice.

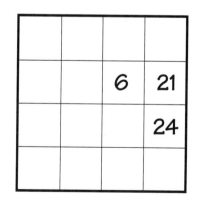

Numbers to place

2	3	5	~~6~~
9	10	12	15
18	20	~~21~~	~~24~~
27	30	35	36

		6	21
			24

Making Connections

Promote reflection and make mathematical connections by asking:

- Which numbers are good beginning numbers? Explain your reasoning.
- What strategies helped the class fill almost all, or all of the cells?

Common Factors Match

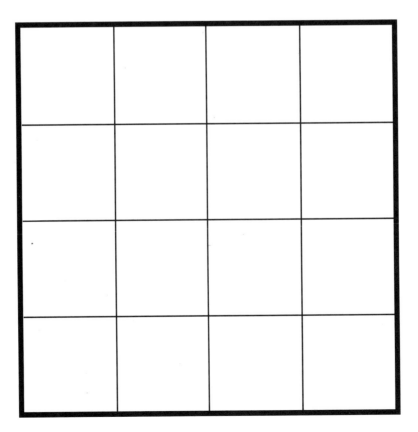

Numbers to place

2	3	5	6
9	10	12	15
18	20	21	24
27	30	35	36

Identifying Values

Topic: Visual Representations of Relationships

Object: Compare the value of unknown symbols.

Groups: Whole class or small group

Materials

- prepared visual problems
- chalkboard or overhead projector

Tip *Save good student-authored problems to share with colleagues and future classes.*

Directions

1. The leader displays both a visual equation which defines two different but equal quantities and a related value problem.

   ❑ = ▲▲

   ❑❑ = ?

2. Students are asked to study the given example and use the defined relationship to help identify the missing equality.
3. When a student volunteers to share a solution, she or he is also required to explain the rationale for her or his answer.
4. The leader provides additional visual clues and problems following these same steps. It is important for students to explain the reasoning used to justify each given solution.

   Additional starters:

   If ▲ ❍ = ▲ ❑, then ❍ = ?

   If ✻ ❍ = ❍❍❑, then ✻ = ?

   If ▲▲ + ☆☆ = ❍❍, then ❍ = ?

   If ▲ + ❍❍ = ■, then ■■ = ?

   If ☆☆☆ = ▲▲▲▲▲▲, then ☆ = ?

   If ■❍❍ = ❍▲▲■, then ❍ = ?

5. After experiencing four puzzles, students should be ready to pair up and create visual clues and related problems for classmates to solve. It is worthwhile to have students write the justification for their solutions.

Making Connections

Promote reflection and make mathematical connections by asking:

- What patterns did you discover that helped you solve these problems?
- What made it challenging to create good puzzles?

What's My Number?

Topic: Representing Numerical Situations Symbolically

Object: Identify an unknown quantity when given a symbolic representation and represent the same number with different symbolic expressions.

Groups: Whole class or small group

Tip Have students work in pairs to prepare and present related numerical situations and symbolic representations for classmates to solve.

Materials

- prepared starter clues
- chalkboard or overhead projector

Directions

1. The leader announces a number relationship and asks a student to represent the relationship with numbers and symbols.

   Example: After the leader states, "Seven more than my number equals 13," student displays $\square + 7 = 13$.

2. After studying the example, students volunteer to give different numerical situations which use the same unknown amount. When a student states a new situation and the leader agrees that the same unknown amount is used, another student records this new relationship.

   Example: A student states, "Doubling my number equals twelve and a different student displays $2n = 12$.

3. Students provide additional descriptions for the same unknown number, as other students represent the stated numerical relationship symbolically.
4. After at least three different equations are shared and displayed, students state the value of the unknown number.
5. The process is repeated with a new unknown number.

   Possible starters:

Four times my number equals 12.	$4n = 12,\ (n = 3)$
Six less than my number equals 19.	$n - 6 = 19,\ (n = 25)$
Five more than double my number equals 15.	$2n + 5 = 15,\ (n = 5)$
Half of my number equals 6.	$\frac{n}{2} = 6,\ (n = 12)$
One less than triple my number equals 20.	$3n - 1 = 20,\ (n = 7)$

6. After three experiences with unknown numbers, students are ready to volunteer similar numerical situations which include an unknown number that can be represented in a variety of ways descriptively and symbolically.

Making Connections

Promote reflection and make mathematical connections by asking:

- Is it easier to create numerical situations or represent situations with symbols and numbers? Please explain.

Date ____________ Name ____________________

Number Explorations 1

STOP Don't start yet. Star problems that may have answers less than 10.

1. Name four factors of 20. ____________

2. Identify the greatest common factor of 12 and 18. _____

3. Name a multiple of 5 between 40 and 50. _____

4. Name a multiple of 8 and 12 between 20 and 30. _____

5. Identify three prime numbers less than 10. ____________

6. $x + 4 = 10$, $x =$ _____ **7.** $3x = 24$, $x =$ _____ **8.** $x \div 3 = 8$, $x =$ _____

9. If $n = 3$, $4n + 5 =$ _____ **10.** If $n = 3$, $20 - 3n =$ _____

Go On Complete the chart.

x	y	$x + y$	$x - y$
30		48	
50			30

Date ____________ Name ____________________

Number Explorations 2

STOP Don't start yet. Star problems that may have odd answers.

1. Name four factors of 16. ____________

2. Identify the greatest common factor of 6 and 14. _____

3. Name a multiple of 7 between 40 and 50. _____

4. Name a multiple of 4 and 6 between 30 and 40. _____

5. Identify two prime numbers between 30 and 40. ____________

6. $30 - x = 24$, $x =$ _____ **7.** $6x = 42$, $x =$ _____ **8.** $30 \div x = 6$, $x =$ _____

9. If $n = 6$, $5n + 4 =$ _____ **10.** If $n = 6$, $40 - 4n =$ _____

Go On My addends are odd, square numbers less than 50. Their sum is 83. What are my 3 numbers? Write another riddle for classmates to solve.

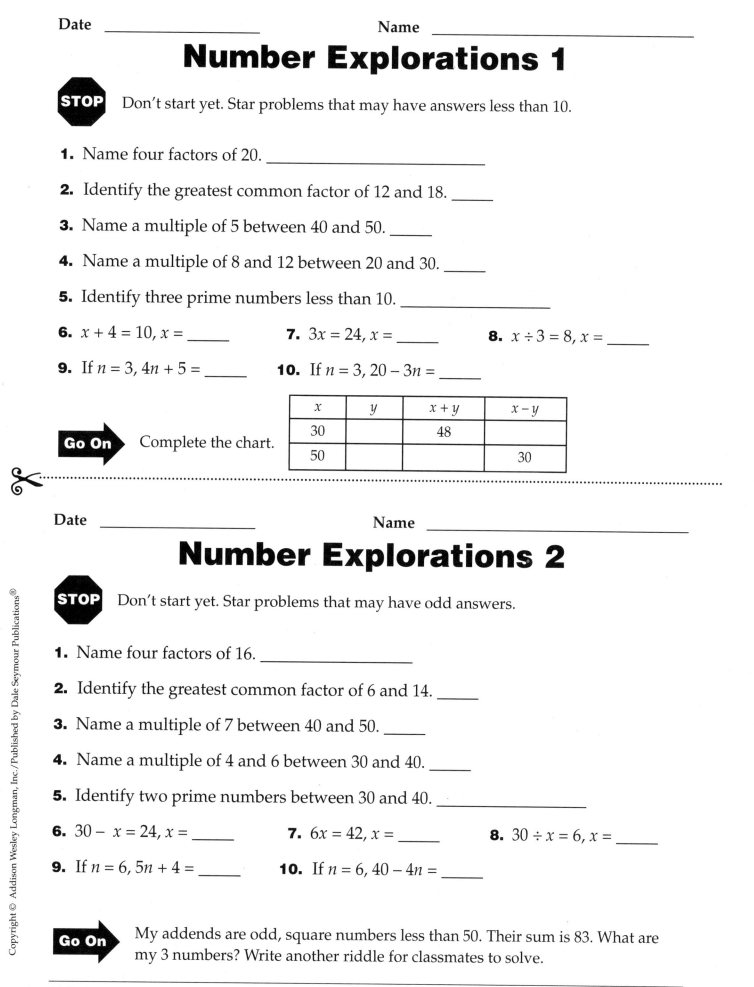

Date ____________ Name ____________________

Number Explorations 3

STOP Don't start yet. Star problems that may have answers between 20 and 30.

1. Name four factors of 18. ________________
2. Identify the greatest common factor of 8 and 12. _____
3. Name a multiple of 6 between 30 and 40. _____
4. Name a multiple of 4 and 3 between 40 and 50. _____
5. Identify two prime numbers between 10 and 15. ____________
6. $x + 7 = 12, x =$ _____
7. $4x = 32, x =$ _____
8. $x \div 3 = 6, x =$ _____
9. If $t = 4$, $6t + 3 =$ _____
10. If $t = 4$, $40 - 5t =$ _____

Go On Complete the chart.

a	b	$a + b$	$a - b$
60		72	
40			15

Date ____________ Name ____________________

Number Explorations 4

STOP Don't start yet. Star problems that may have answers that are multiples of 5.

1. Name four factors of 24. ________________
2. Identify the greatest common factor of 10 and 16. _____
3. Name a multiple of 8 between 40 and 60. _____
4. Name a multiple of 3 and 15 between 40 and 50. _____
5. Identify two prime numbers between 15 and 20. ____________
6. $27 - x = 17, x =$ _____
7. $6x = 48, x =$ _____
8. $36 \div x = 9, x =$ _____
9. If $t = 7$, $4t + 6 =$ _____
10. If $t = 7$, $50 - 7t =$ _____

Go On My numbers are multiples of 6, greater than 25 and less than 50. Their sum is 120. What are my 3 numbers? Write another riddle for classmates to solve.

Date ____________ Name ____________

Number Explorations 5

STOP Don't start yet. Star problems that may have answers greater than 30.

1. Name four factors of 12. ____________
2. Identify the greatest common factor of 15 and 18. _____
3. Name a multiple of 4 between 20 and 30. _____
4. Name a multiple of 2 and 6 between 20 and 25. _____
5. Identify two prime numbers between 20 and 30. ____________
6. $x + 8 = 11$, $x =$ _____
7. $4x = 20$, $x =$ _____
8. $x \div 2 = 8$, $x =$ _____
9. If $c = 5$, $8c + 8 =$ _____
10. If $c = 5$, $9c - 4 =$ _____

Go On Complete the chart.

c	d	$c + d$	$c - d$
80		95	
60			30

Date ____________ Name ____________

Number Explorations 6

STOP Don't start yet. Star problems that may have even answers.

1. Name four factors of 36. ____________
2. Identify the greatest common factor of 8 and 16. _____
3. Name a multiple of 9 between 30 and 50. _____
4. Name a multiple of 8 and 6 between 70 and 80. _____
5. Identify three prime numbers between 40 and 50. ____________
6. $35 - x = 29$, $x =$ _____
7. $5x = 35$, $x =$ _____
8. $36 \div x = 6$, $x =$ _____
9. If $c = 8$, $4c + 5 =$ _____
10. If $c = 8$, $6c - 5 =$ _____

Go On My numbers are multiples of 4 and 6, and less than 50. Their sum is 84. What are my 3 numbers? Write another riddle for classmates to solve.

Greatest Common Factor Bingo

Topic: Greatest Common Factor

Object: Cover four numbers in a row.

Groups: 2 players or pair players

Materials for each group

- *Greatest Common Factor Bingo* gameboard, p. 143
- markers (different kind for each player)
- pencil and 2 paper clips (for spinners)
- scratch paper

Tips Explore how this game differs if players are allowed to replace opponent's markers that cover corresponding greatest common factors. If preferred, the blank number cubes, p. 153, can be substituted for the spinner.

Directions

1. The first player uses the pencil and paper clips with the two spinners to identify and announce two composite numbers.
2. The second player determines the greatest common factor for the two numbers spun and diagrams the two corresponding factor trees to provide proof. (The greatest common factor can be found by multiplying the common factors in the bottom rows.)

   Example: 12 and 30 are spun. The player states, "Six is the greatest common factor," and displays factor tree proof.

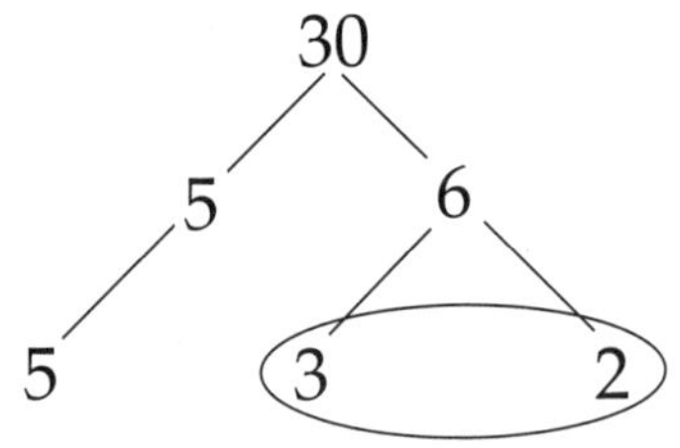

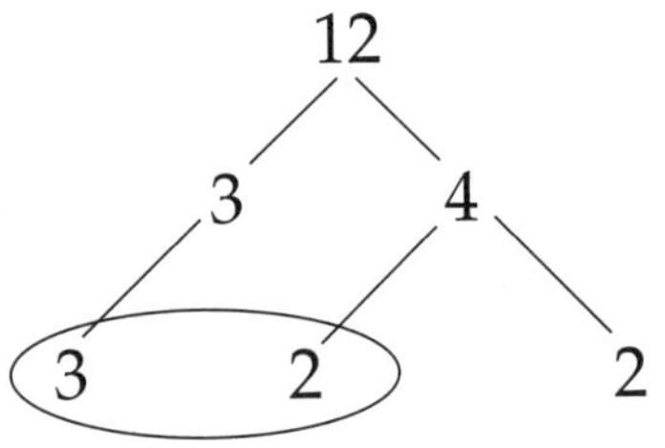

3. Once the first player accepts the second player's response, the second player covers the resulting greatest common factor. If the greatest common factor appears twice on the gameboard, only one may be covered each turn.
4. The second player spins and announces the two spun numbers. After the first player identifies the greatest common factor for the two numbers and provides correct factor trees, the player covers a corresponding cell on the gameboard.
5. If the resulting greatest common factor is already covered by the opponent's marker, the player is allowed one additional spin of only one of the spinners.
6. This procedure and alternating of turns continues until one player positions four markers in a row horizontally, vertically, or diagonally.

Making Connections

Promote reflection and make mathematical connections by asking:

- What approach helped you quickly identify the greatest common factor?
- How could this game be modified to promote more mathematical thinking?

Greatest Common Factor Bingo

2	6	10	20	4
4	8	4	2	12
12	2	6	8	4
10	4	8	4	2
2	16	2	6	4

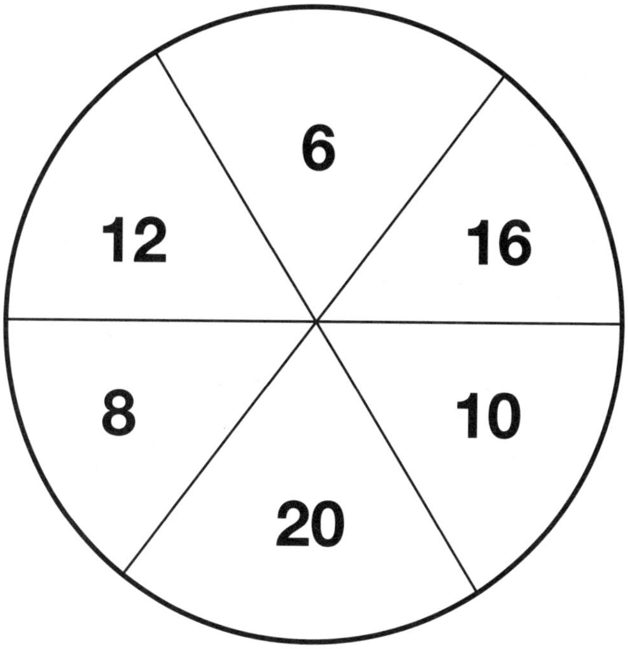

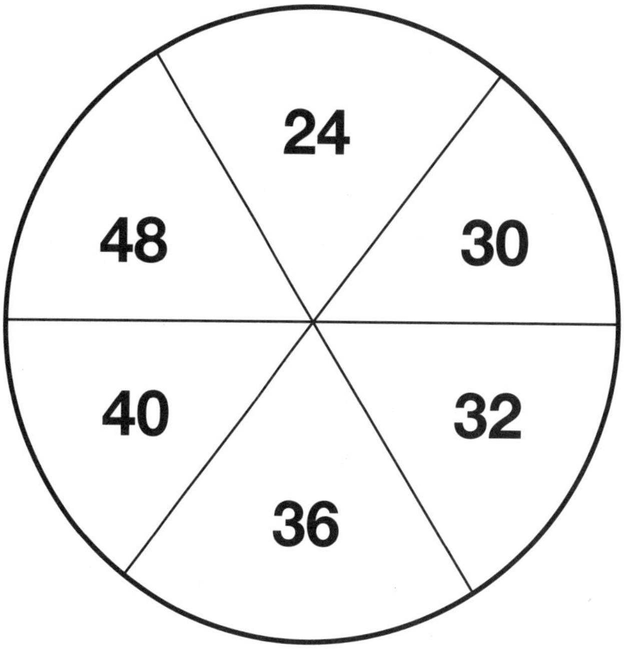

Sixty

Topic: Number Theory

Object: Accumulate points by identifying primes, multiples, factors and perfect squares.

Groups: 2 players or pair players

Materials for each group

- 2 sets of Digit Cards (0 removed), p. 150
- scratch paper for tallying scores

Tips *Play two or more rounds and identify what factors contribute to high-scoring games. Some students enjoy playing this game until the accumulated total exceeds 100. (An additional set of Digit Cards will be needed.)*

Directions

1. After removing the zeros, mix two sets of Digit Cards and stack them face down.
2. The first player turns over the top card and announces the number on the card which becomes the current total. This card is displayed face-up and begins the discard pile.
3. Players receive one point each if the current total equals a prime number, a multiple of 3 or 4 , a factor of 60 and/or a square number. However, the first player is only allowed one point for her or his first turn.

   Example: The first player displays 3. The player qualifies for one point since 3 can qualify as a prime, a multiple of 3, or a factor of 60.
4. The second player turns over the next card, adding its value to the discard pile's previous total. After the player states the new total, the player explains any points earned and tallies the points under her or his name.

   Example: Player two displays 6 and announces nine as the new total. The player tallies two points if nine is recognized as a square number and as a multiple of three.
5. The first player turns over the next card, places it face-up in the discard pile, and announces the new total.
6. After the player identifies and explains about any earned points, the player tallies the earned points.
7. Players continue to alternate turns following these steps until the accumulated discard total reaches or exceeds 60. At this point the player with the highest tallied score wins the round.

Making Connections

Promote reflection and make mathematical connections by asking:

- What helped you recognize when totals qualified for more than one point?
- How might it be possible to receive three points on a turn?

Seek and Cover

Tip *Provide additional practice by having students create more challenging clues.*

Topic: Solving for One Unknown

Object: Score the most points by aligning similar markers in 3 or 4-in-a-row arrangements.

Groups: 2 players or pair players

Materials for each group

- *Seek and Cover* gameboard, p. 146
- special number cube (2, 3, 4, 5, 6, Choose), p. 153
- different kind of markers for each player (8 each)

Directions

1. The first player rolls the number cube and locates cells that display an equation where the variable equals the number rolled. If the player rolls "Choose," any cell may be covered but the player needs to identify the value of the variable. After the player selects and covers one of the cells, the player justifies that her or his solution is correct.
2. The second player follows the same process of rolling the number cube, selecting and covering a cell, and justifying this action aloud.
3. Players continue to alternate turns and follow these steps. If the rolled number does not fit any uncovered cell, the player passes and does not place a marker on that turn.
4. Play continues until both players have all sixteen markers placed or either player passes three consecutive times.
5. Each player determines her or his final score by scoring one point for every three-in-a-row and three points for every four-in-a-row arrangement. Rows can be horizontal, vertical or diagonal.
6. At the end of the game, each player demonstrates how her or his specified score was achieved.

Making Connections

Promote reflection and make mathematical connections by asking:

- What approaches seem to generate higher totals?
- How will you play this game differently next time?

Seek and Cover

$5n = 10$	$\frac{15}{x} = 3$	$9 - x = 5$	$7 - x = 4$
$n - 2 = 4$	$8 + n = 12$	$4n = 12$	$6 + n = 12$
$3n = 15$	$9 + x = 11$	$n - 3 = 2$	$2n = 8$
$\frac{12}{x} = 3$	$3n = 18$	$x + 5 = 8$	$\frac{8}{x} = 4$

Date ____________________ Name ______________________________

Number Riddle Arrangements

Use the row and column clues to place the displayed numbers in each puzzle. Make sure all clues are satisfied. (The cell in the first row and first column is designated by heavier border lines.)

Row Clues:
Two odd numbers belong in the second row.
The sum of the first row numbers equals a prime number.

1	3	7	10

Column Clues:
The sum of the numbers in the first column is the greatest common factor of 16 and 24.
The difference between the numbers in the second column is a prime > 5.

Row Clues:
The sum of the numbers in the second row equals 12.
The difference of the two numbers in first row is an even prime number.

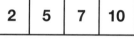

2	5	7	10

Column Clues:
The sum of the numbers in each column is a multiple of 3.
The difference of the numbers in each column equals the same prime number.

Row Clues:
The sum of the numbers in the first row is a multiple of 7.
The sum of the numbers in the second row is a multiple of 2 and 3.

1	2	4	5	6	8

Column Clues:
The sum of the numbers in the first column is a prime number.
The product of the numbers in the third column is a square number.
The product of the numbers in the second column is a multiple of 10.

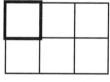

Row Clues:
The sum of the numbers in the first row is a square number.
Each number in the second row is divisible by 3.

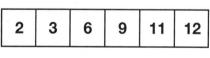

2	3	6	9	11	12

Column Clues:
The differences between the two numbers in the first column and the two numbers in the second column are prime numbers.
The product of the numbers in the third column is a square number.
The product of the numbers in the third column is twice the product of the numbers in the second column.

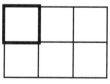

Work with a classmate to create a similar number riddle for other classmates to solve.

Date ____________________ Name ________________________________

Discover Trivia by Identifying Unknowns

Solve each equation.

S $7 + x = 18$
$x =$

U $2n = 8$
$n =$

L $x + 5 = 14$
$x =$

O $4 + n = 17$
$n =$

V $x - 9 = 6$
$x =$

I $15 - n = 8$
$n =$

N $2x + x = 24$
$x =$

I $n - 9 = 3$
$n =$

N $3x = 18$
$x =$

A $16 \div n = 8$
$n =$

I $3x = 15$
$x =$

E $n \div 2 = 8$
$n =$

E $3x = 30$
$x =$

L $n \div 2 = 7$
$n =$

To identify the most landed-on square in Monopoly®, locate each equation's solution below and record the corresponding letter of the problem on each blank.

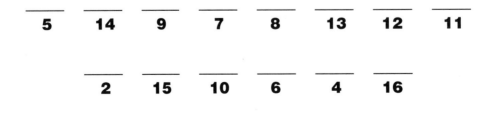

___	___	___	___	___	___	___	___
5	**14**	**9**	**7**	**8**	**13**	**12**	**11**

___	___	___	___	___	___
2	**15**	**10**	**6**	**4**	**16**

Monopoly®, is a registered trademark of Parker Brothers division of Tonka Corporation.

Blackline Masters

Digit Cards

Digit Squares

Multiplication Chart

Number Cubes (1–6, blank)

Number Cubes (4–9), (.1–.6)

Reach the Peak

Spinners (1–6), blank

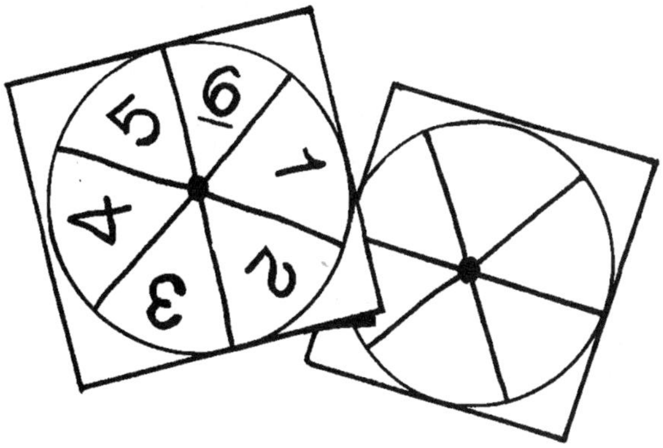

Digit Cards

0		
1	2	3
4	5	6
7	8	9

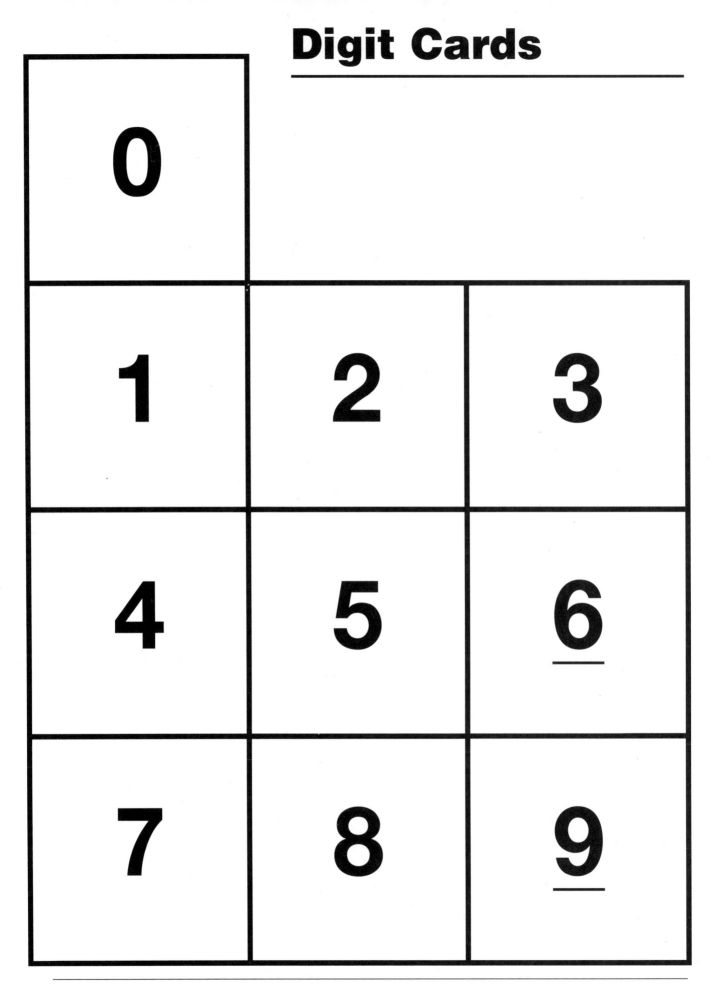

Digit Squares

0	1	2	3	4
5	6	7	8	9

0	1	2	3	4
5	6	7	8	9

0	1	2	3	4
5	6	7	8	9

Multiplication Chart

x	1	2	3	4	5	6	7	8	9	10
1	1	2	3	4	5	6	7	8	9	10
2	2	4	6	8	10	12	14	16	18	20
3	3	6	9	12	15	18	21	24	27	30
4	4	8	12	16	20	24	28	32	36	40
5	5	10	15	20	25	30	35	40	45	50
6	6	12	18	24	30	36	42	48	54	60
7	7	14	21	28	35	42	49	56	63	70
8	8	16	24	32	40	48	56	64	72	80
9	9	18	27	36	45	54	63	72	81	90
10	10	20	30	40	50	60	70	80	90	100

Number Cubes

Cut solid lines. Fold on dotted lines.

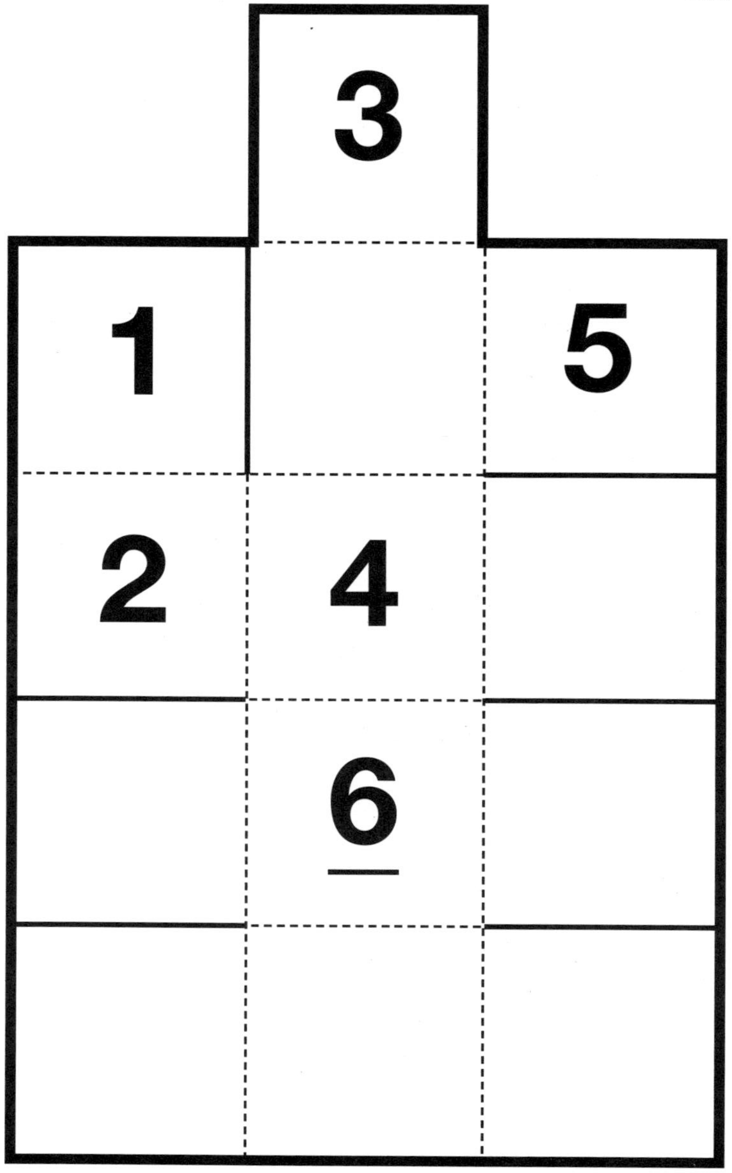

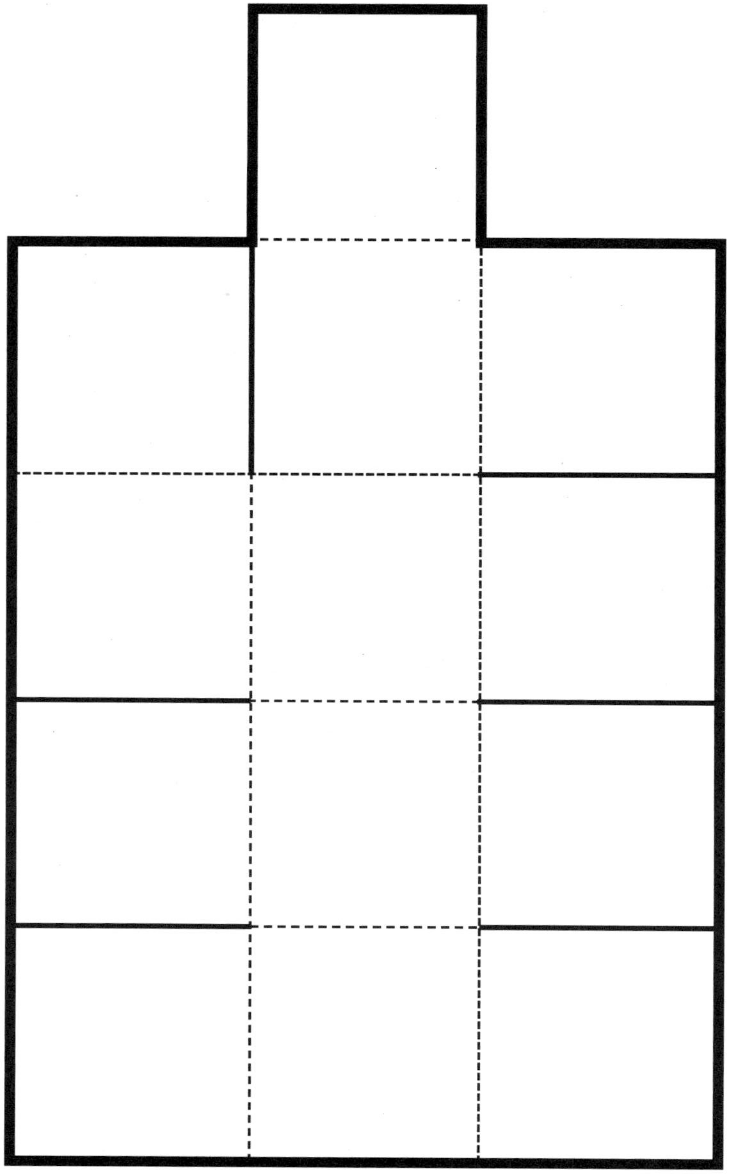

Number Cubes

Cut solid lines. Fold on dotted lines.

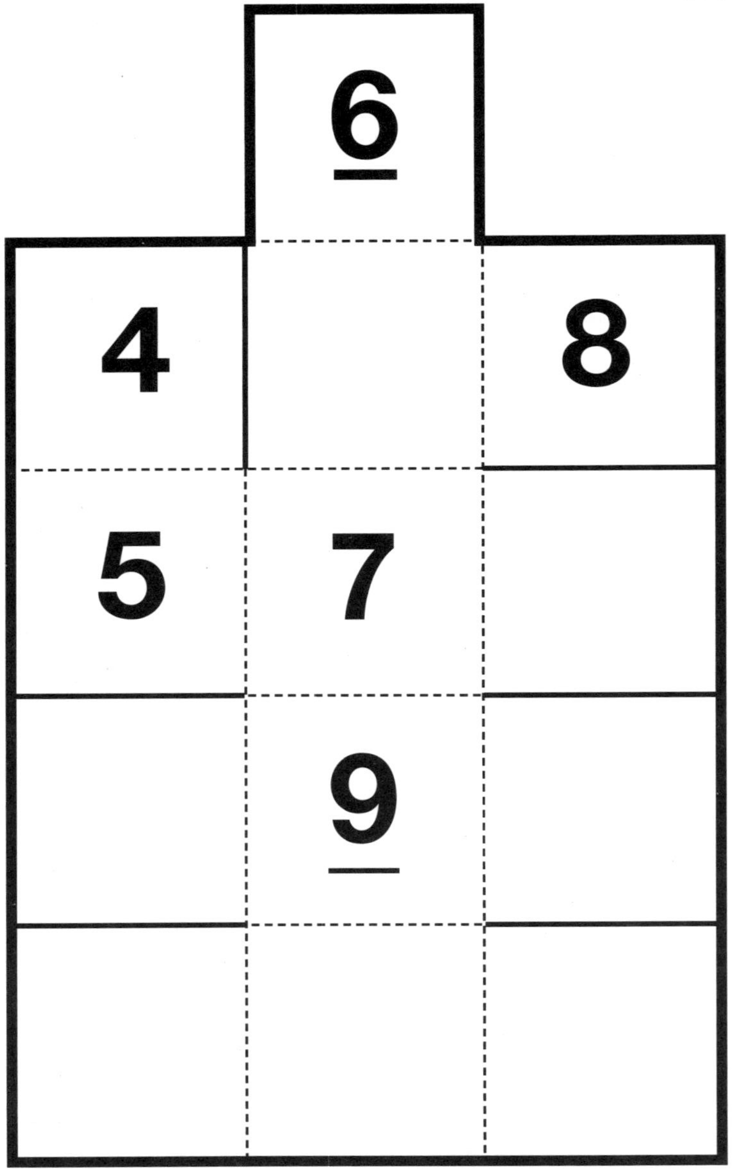

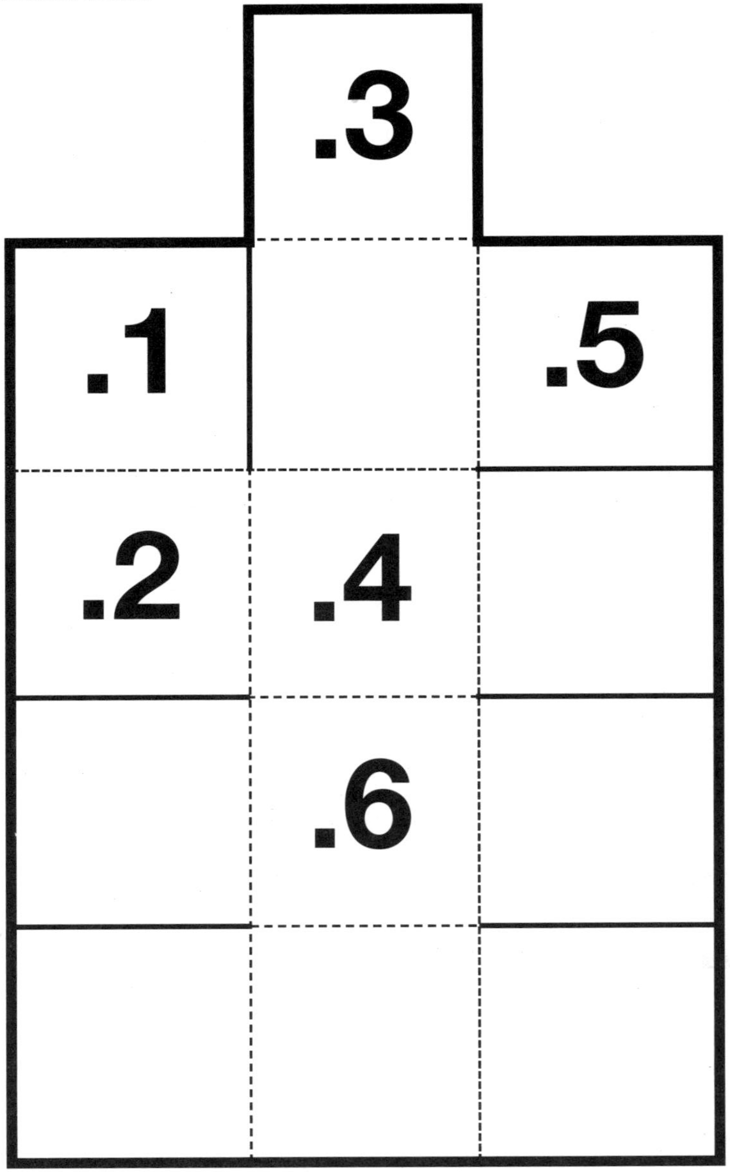

Reach the Peak

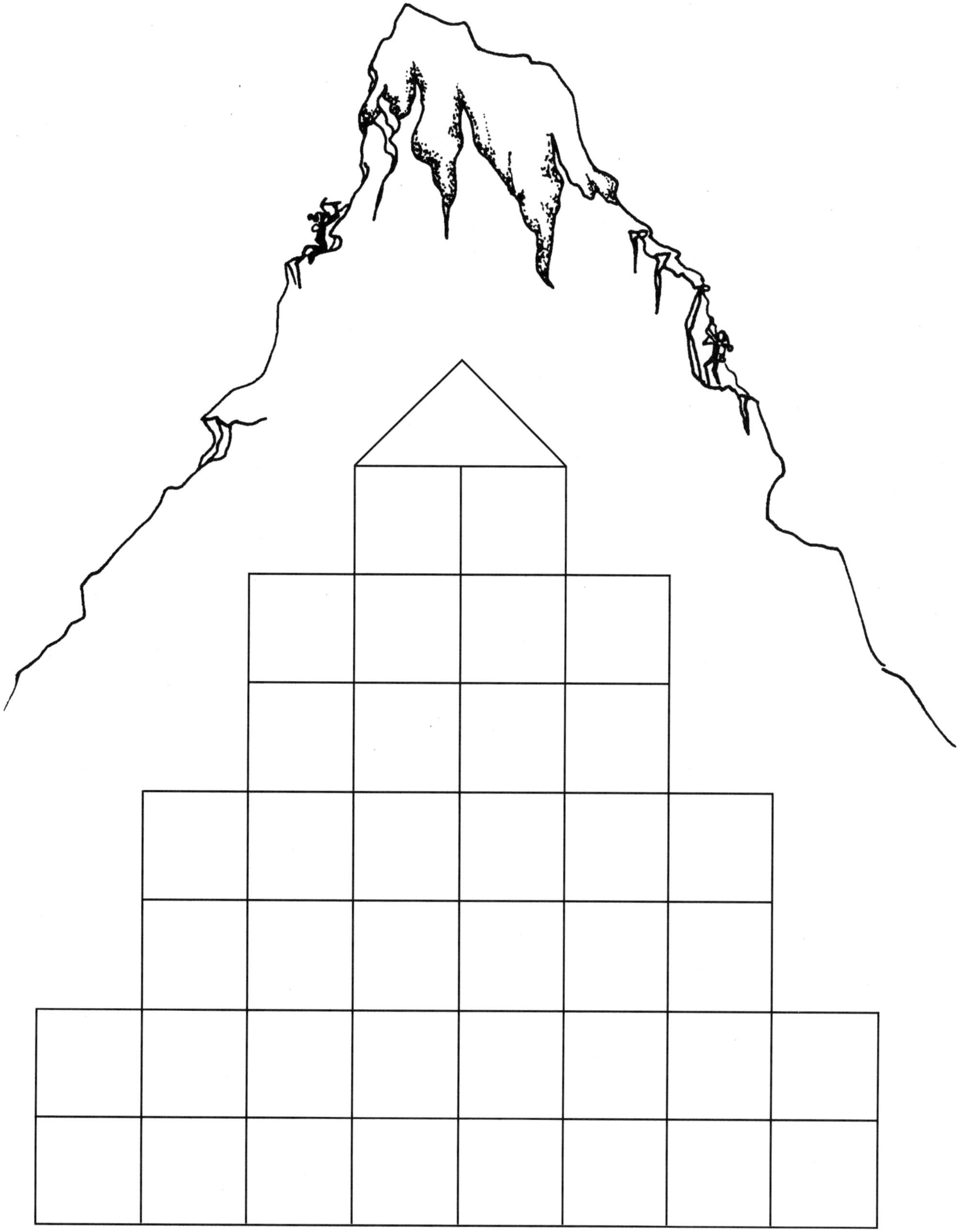

Spinners

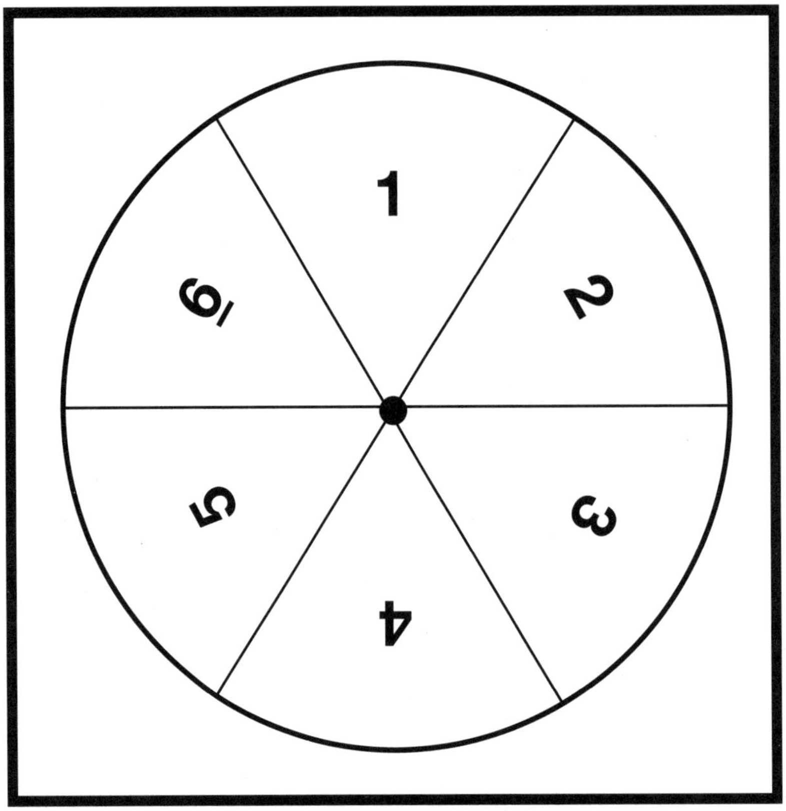

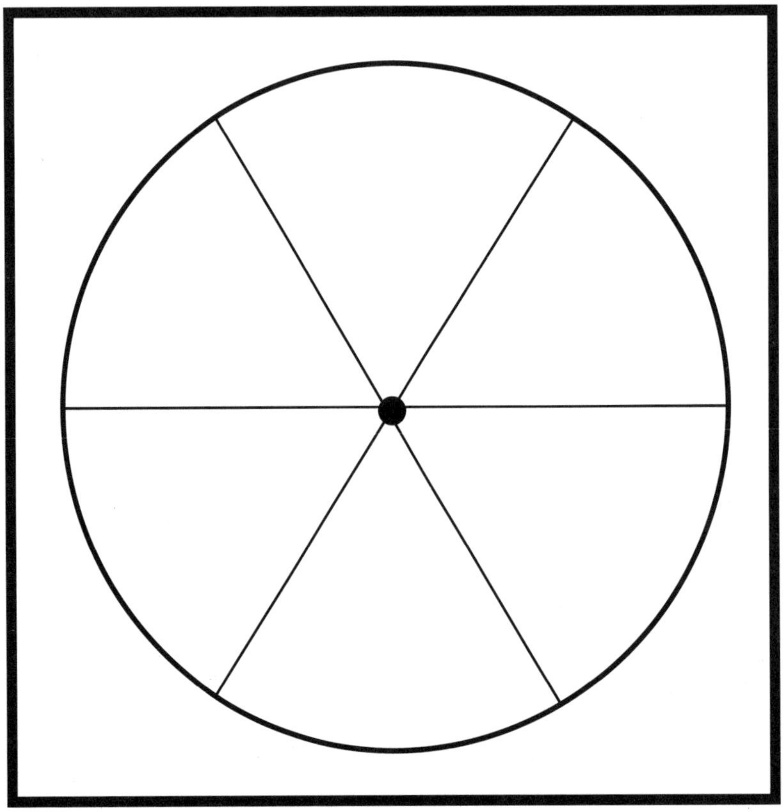

Nimble with Numbers Answer Key

p. 17 *Quick Check 1* For problems 4–6, the order of addends and/or factors may vary.
1) 34 2) 36 3) 25 4) (8 × 4) – 9 5) 8 × 3 ÷ 4 6) (6 + 3) × (8 ÷ 2) 7) (4 ÷ 2) × (9 – 3)
8) 6 × 4 ÷ 3 9) (9 + 7) ÷ 4 10) 4 × (9 – 7) + 4 or (4 × 9) ÷ (7 – 4)
Go On Answers will vary.

Quick Check 2 For problems 4–6, the order of addends and/or factors may vary.
1) 38 2) 48 3) 32 4) (7 × 6) – 8 5) 9 × 2 ÷ 6 6) (6 + 4) × (9 ÷ 3) 7) (6 ÷ 3) × (9 – 1)
8) (9 ÷ 3) + 6 9) (9 + 7) ÷ 8 10) (5 + 4) × (12 ÷ 4)
Go On Answers will vary.

p. 18 *Quick Check 3* For problems 4–6, the order of addends and/or factors may vary.
1) 17 2) 35 3) 24 4) (6 × 5) – 8 5) 6 × 4 ÷ 3 6) (2 + 7) × (8 ÷ 4) 7) (6 ÷ 2) × (8 – 1)
8) 8 × 6 ÷ 4 9) (4 + 3) × 3 10) (7 + 5) ÷ (9 ÷ 3)
Go On Answers will vary.

Quick Check 4 For problems 4–6, the order of addends and/or factors may vary.
1) 54 2) 63 3) 24 4) (3 × 9) – 7 5) 2 × 6 ÷ 4 6) (3 + 4) × (6 ÷ 2) 7) (9 ÷ 3) × (8 – 1)
8) 8 × 3 ÷ 6 9) (5 + 9) ÷ 2 10) (6 × 3) + 4 + 7
Go On Answers will vary.

p. 19 *Quick Check 5* For problems 4–6, the order of addends and/or factors may vary.
1) 50 2) 54 3) 25 4) (8 × 7) – 9 5) 3 × 6 ÷ 9 6) (4 + 5) × (6 ÷ 3) 7) (8 ÷ 2) × (9 – 6)
8) 6 × 3 + 4 9) 6 + (12 ÷ 4) 10) (9 + 3) ÷ (8 – 5)
Go On 36, 30 Pattern: triple the number, subtract 6

Quick Check 6 For problems 4–6, the order of addends and/or factors may vary.
1) 39 2) 42 3) 18 4) (4 × 9) – 7 5) 3 × 4 ÷ 2 6) (5 + 7) × (4 ÷ 2) 7) (9 ÷ 3) × (7 – 4)
8) (9 × 3) – 5 9) 8 – 7 +3 10) (8 × 4) – (6 × 4) or (8 – 4) × (6 – 4)
Go On Answers will vary.

p. 27 ***Valuable Equations:*** Answers will vary.

p. 28 *Joining Neighbors*
Answers will vary. Samples are given.

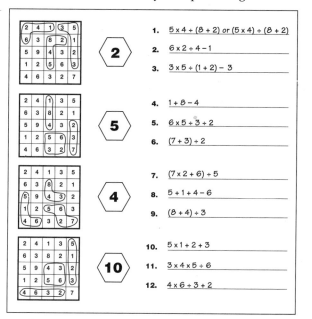

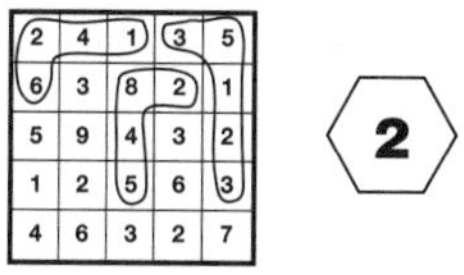

1. 5 x 4 ÷ (8 + 2) or (5 x 4) ÷ (8 + 2)
2. 6 x 2 ÷ 4 – 1
3. 3 x 5 ÷ (1 + 2) – 3

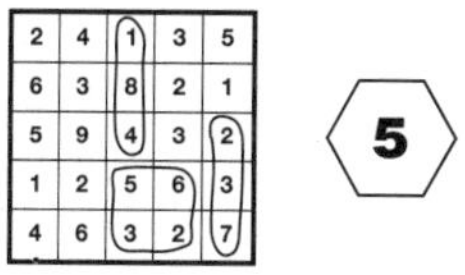

4. 1 + 8 – 4
5. 6 x 5 ÷ 3 ÷ 2
6. (7 + 3) ÷ 2

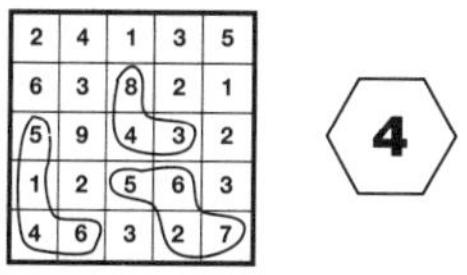

7. (7 x 2 + 6) ÷ 5
8. 5 + 1 + 4 – 6
9. (8 + 4) ÷ 3

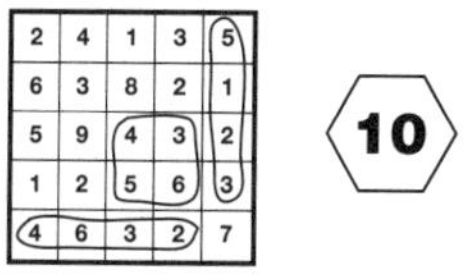

10. 5 x 1 + 2 + 3
11. 3 x 4 x 5 ÷ 6
12. 4 x 6 ÷ 3 + 2

p. 29 *Joining Neighbors II*
Answers will vary. Samples are given.

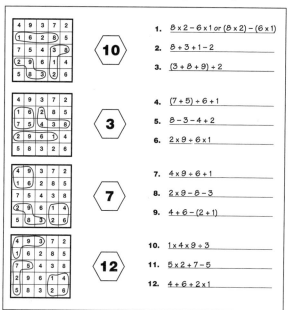

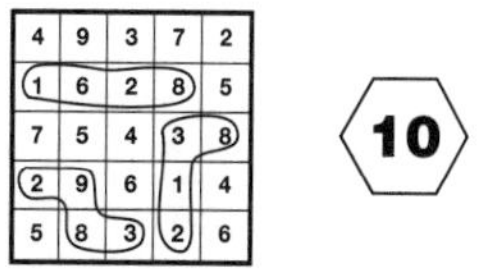

1. 8 x 2 – 6 x 1 or (8 x 2) – (6 x 1)
2. 8 + 3 + 1 – 2
3. (3 + 8 + 9) ÷ 2

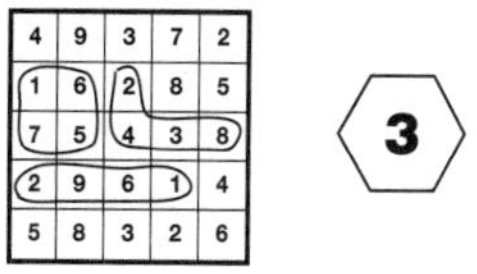

4. (7 + 5) ÷ 6 + 1
5. 8 – 3 – 4 + 2
6. 2 x 9 ÷ 6 x 1

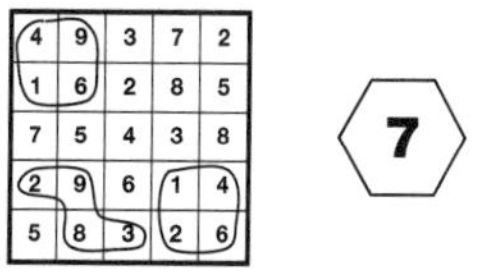

7. 4 x 9 ÷ 6 + 1
8. 2 x 9 – 8 – 3
9. 4 + 6 – (2 + 1)

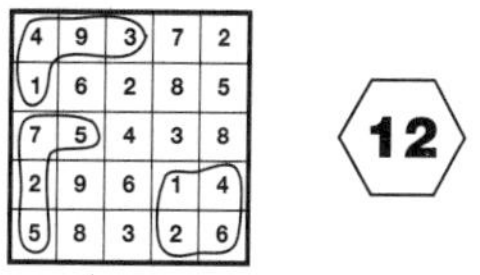

10. 1 x 4 x 9 ÷ 3
11. 5 x 2 + 7 – 5
12. 4 + 6 + 2 x 1

p. 30 ***Matches Count Practice***
Answers will vary. Samples are given

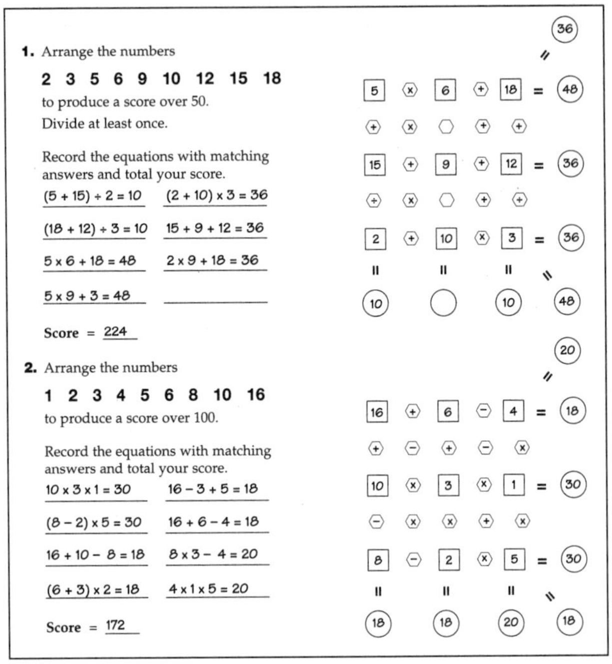
1. Arrange the numbers
2 3 5 6 9 10 12 15 18
to produce a score over 50.
Divide at least once.

Record the equations with matching answers and total your score.

(5 + 15) ÷ 2 = 10	(2 + 10) × 3 = 36
(18 + 12) ÷ 3 = 10	15 + 9 + 12 = 36
5 × 6 + 18 = 48	2 × 9 + 18 = 36
5 × 9 + 3 = 48	

Score = 224

2. Arrange the numbers
1 2 3 4 5 6 8 10 16
to produce a score over 100.

Record the equations with matching answers and total your score.

10 × 3 × 1 = 30	16 − 3 + 5 = 18
(8 − 2) × 5 = 30	16 + 6 − 4 = 18
16 + 10 − 8 = 18	8 × 3 − 4 = 20
(6 + 3) × 2 = 18	4 × 1 × 5 = 20

Score = 172

p. 39 ***In Your Head 1***
1) 157 2) 138 3) 394 4) 840 5) 64 6) 162 7) 4200
8) 870 9) 12*r*5 10) 77
Go On ×, ÷; −, ×

In Your Head 2
1) 172 2) 317 3) 636 4) 1600 5) 144 6) 252 7) 20,000
8) 2450 9) 13*r*3 10) 87
Go On 83 × 64 = 5312

p. 40 ***In Your Head 3***
1) 143 2) 226 3) 487 4) 600 5) 85 6) 144 7) 5600
8) 760 9) 14*r*4 10) 84
Go On 245 − 76 = 169

In Your Head 4
1) 176 2) 342 3) 772 4) 2400 5) 171 6) 266 7) 48,000
8) 2340 9) 13*r*3 10) 76
Go On ÷, −; ÷, −

p. 41 ***In Your Head 5***
1) 149 2) 263 3) 626 4) 720 5) 102 6) 315 7) 72,000
8) 2070 9) 14*r*1 10) 65
Go On 356 × 2 = 712

In Your Head 6
1) 185 2) 355 3) 879 4) 960 5) 126 6) 352 7) 30,000
8) 2950 9) 16*r*2 10) 72
Go On 764 ÷ 2 = 382

p. 49 ***Rearrange and Find I***
1) 41 − 5 2) 45 × 2 3) 54 ÷ 2 4) 41 × 2 5) 52 × 4 6) 52 ÷ 4
7) 41 − 25 8) 25 × (4 + 1) 9) 12 + (5 − 4) 10) (42 × 5) − 1 11) 152 × 4 12) 214 × 5
13) 62 − 4 14) 26 × 3 15) 42 ÷ 3 16) 36 + 4 17) 32 × 6 or 64 × 3
18) 43 × 6 19) 62 − 34 20) (26 × 4) + 3 21) 26 × (4 + 3) 22) (42 − 3) × 6
23) 362 × 4 24) 432 × 6

p. 50 ***Rearrange and Find II***

1) 69 + 5 or 65 + 9 2) 39 × 5 or 65 × 3 3) 96 ÷ 3 4) 93 – 6 5) 69 ÷ 3 6) 63 × 9
7) 93 – 56 8) (35 + 9) × 6 or (53 – 9) × 6 9) (93 – 6) × 5 10) 35 × 9 – 6 or 53 × 6 – 9
11) 96 + 35 or 95 + 36 12) 35 × (9 – 6) 13) 935 × 6 14) 395 × 6 15) 27 × 4
16) 87 + 4 or 84 + 7 17) 74 × 8 18) 84 – 7 19) 78 ÷ 2 20) 72 × 4
21) 47 + 28 or 27 + 48 22) 48 × 2 + 7 23) (74 – 8) ÷ 2 24) 82 + 4 × 7
25) 27 × 8 ÷ 4 or 82 – 4 × 7 26) 72 + 4 + 8 or 28 × (7 – 4) 27) 824 – 7 28) 287 × 4

p. 51 ***Related Cross-Number Puzzle***

1. 1	2. 9	8			3. 4	0	4. 5
5. 8	1		6. 3	6	0		4
	7. 2	2	0		8. 8	9. 7	5
10. 2			11. 4	12. 8	8	2	
13. 3	14. 5			2			15. 3
16. 8	6	17. 5		18. 2	19. 4		6
	20. 9	5	8		21. 1	2	9

p. 52 ***Operation Fill Practice***

1) (57 + 128) ÷ 5 2) 74 + 69 – 25 3) (938 – 98) ÷ 4 4) (81 – 75) × 32
5) 6 × 35 – 146 6) 324 ÷ 3 + 42 7) 192 ÷ 8 × 12 8) 6 × 47 + 8 – 46
9) 9 × 210 ÷ 3 ÷ 2 10) 39 × 8 – 57 + 69 11) (102 ÷ 6 + 43) ÷ 5

p. 57 ***Product and Quotients Galore 1***

1) 4536 2) 52,640 3) 864 4) 3915 5) 14,420 6) 37*r*2
7) 7*r*14 8) 76*r*6 9) 57 10) 48*r*2 Go On 743 × 86 = 63, 898

Product and Quotients Galore 2

1) 7160 2) 24,960 3) 1768 4) 6566 5) 22,618 6) 45*r*4
7) 8*r*15 8) 82*r*4 9) 64 10) 53*r*1 Go On 856 ÷ 4 = 214

p. 58 ***Product and Quotients Galore 3***

1) 7083 2) 32,220 3) 1505 4) 3002 5) 19,575 6) 52*r*3
7) 7*r*11 8) 53*r*3 9) 76 10) 74*r*1 Go On ×, –, ÷; ÷, +, ÷

Product and Quotients Galore 4

1) 6692 2) 69,920 3) 1591 4) 6764 5) 29,052 6) 64*r*2
7) 9*r*13 8) 64*r*7 9) 62 10) 65*r*2 Go On 3569 × 2 = 7138

p. 59 ***Product and Quotients Galore 5***

1) 6352 2) 42,300 3) 1350 4) 3666 5) 11,340 6) 58*r*3
7) 8*r*12 8) 57*r*6 9) 73 10) 72*r*2 Go On 539 ÷ 7 = 77

Product and Quotients Galore 6

1) 7794 2) 59,310 3) 1980 4) 8633 5) 33,644 6) 66*r*5
7) 9*r*16 8) 68*r*8 9) 58 10) 57*r*1 Go On +, ÷, +; +, ×, ÷

p. 66 ***Detect the Digits***

1) 3456 2) 675; 4725 3) 6; 4974 4) 40; 18, 280 5) 532 6) 459; 1836
7) 476; 3808 8) 53; 1590 9) 13; 338 10) 30; 2670 11) 74; 3034 12) 533; 11, 193
13) 45; 540 14) 29; 192; 5568 15) 63; 29; 1827 16) 84; 4297

p. 67 *Products Galore I*

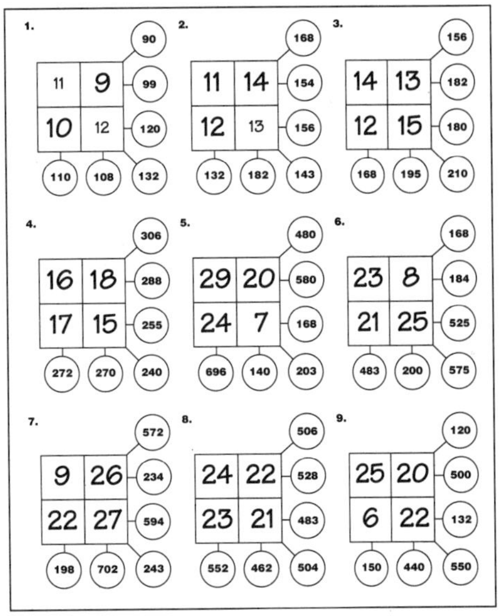

p. 68 *Products Galore II*

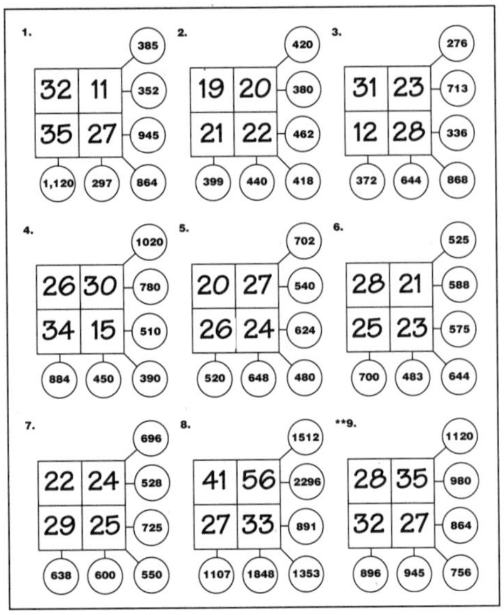

p. 69 *Finding Products Practice*

1) 19 × 38 = 722, 19 × 42 = 798, or 21 × 38 = 798
2) 19 × 21 = 399, 19 × 85 = 1615, or 21 × 85 = 1785
3) 21 × 85 = 1785, 38 × 42 = 1596, 42 × 70 = 2940, 42 × 85 = 3570, or 21 × 70 = 1470
4) 19 × 70 = 1330 or 21 × 70 = 1470
5) 19 × 38 = 722
6) 70 × 85 = 5950
7) 63 × 98 = 6174
8) 17 × 98 = 1666, 34 × 49 = 1666, or 34 × 56 = 1904
9) 17 × 49 = 833
10) 49 × 63 = 3087
11) 22 × 98 = 2156 or 34 × 63 = 2142
12) 17 × 22 = 374

p. 70 *Finding Quotients Practice*

1) 180 ÷ 60 = 3 or 180 ÷ 36 = 5
2) 2580 ÷ 15 = 172
3) 2580 ÷ 60 = 43, 1081 ÷ 23 = 47, 2052 ÷ 36 = 57, or 915 ÷ 15 = 61
4) 1081 ÷ 60 = 18*r*1 or 915 ÷ 60 = 15*r*15
5) 180 ÷ 23 = 7*r*19, 915 ÷ 60 = 15*r*15, 915 ÷ 23 = 39*r*18, 915 ÷ 36 = 25*r*15, or 2580 ÷ 36 = 71*r*24
6) 1081 ÷ 15 = 72*r*1, 1081 ÷ 26 = 30*r*1, or 1081 ÷ 60 = 18*r*1
7) 486 ÷ 54 = 9
8) 486 ÷ 54 = 9, 594 ÷ 54 = 11, 486 ÷18 = 27, or 594 ÷ 18 = 33
9) 594 ÷ 39 = 15*r*9, 594 ÷ 45 = 13*r*9, 1080 ÷ 39 = 27*r*27, 2052 ÷ 45 = 45*r*27, or 1248 ÷ 45 = 27*r*33
10) 486 ÷ 18 = 27, 1248 ÷ 39 = 32, or 594 ÷ 18 = 33
11) 1080 ÷ 39 = 27*r*27, 2052 ÷ 45 = 45*r*27, 1248 ÷ 45 = 27*r*33, or 486 ÷ 45 = 10*r*36
12) 1248 ÷ 18 = 69*r*6

p. 78 *Partial Possibilities 1*

1) $\frac{1}{3}, \frac{4}{9}, \frac{1}{2}, \frac{5}{6}$ 2) $\frac{7}{10}$ 3) $\frac{1}{2}$ 4) $\frac{11}{12}$ 5) $1\frac{1}{2}$ 6) $1\frac{4}{7}$

7) $1\frac{3}{4}$ 8) $\frac{1}{6}$ 9) $\frac{3}{10}$ 10) 16

Go On $2\frac{11}{12}, 3\frac{1}{12}, 3\frac{1}{4}$; numbers increase by $\frac{1}{6}$

Partial Possibilities 2

1) $\frac{1}{2}, \frac{2}{3}, \frac{3}{4}, \frac{5}{6}$ 2) $1\frac{1}{9}$ 3) $\frac{1}{12}$ 4) $\frac{13}{20}$ 5) $2\frac{5}{6}$ 6) $1\frac{1}{6}$

7) $2\frac{1}{12}$ 8) $\frac{7}{12}$ 9) $\frac{1}{3}$ 10) $1\frac{2}{3}$

Go On Answers will vary

p. 79 ***Partial Possibilities 3***

1) $\frac{1}{3}, \frac{7}{12}, \frac{3}{4}, \frac{5}{6}$ 2) $\frac{7}{12}$ 3) $\frac{3}{8}$ 4) $\frac{23}{24}$ 5) $1\frac{5}{8}$ 6) $1\frac{3}{5}$
7) $1\frac{7}{8}$ 8) $\frac{4}{15}$ 9) $\frac{2}{5}$ 10) $2\frac{2}{3}$

Go On Some possible answers $\frac{6}{16}$ (not in lowest terms); $\frac{8}{9}$ (not close to $\frac{1}{2}$)

Partial Possibilities 4

1) $\frac{1}{4}, \frac{3}{10}, \frac{2}{5}, \frac{1}{2}$ 2) $\frac{7}{8}$ 3) $\frac{1}{12}$ 4) $\frac{14}{15}$ 5) $2\frac{3}{4}$ 6) $1\frac{5}{8}$
7) $2\frac{1}{10}$ 8) $\frac{1}{24}$ 9) $\frac{1}{4}$ 10) 21

Go On $3\frac{8}{9}, 4\frac{4}{9}$, 5; numbers increase by $\frac{5}{9}$

p. 80 ***Partial Possibilities 5***

1) $\frac{1}{2}, \frac{5}{6}, \frac{7}{8}, \frac{11}{12}$ 2) $\frac{5}{6}$ 3) $\frac{1}{3}$ 4) $\frac{11}{12}$ 5) $1\frac{8}{9}$ 6) $1\frac{5}{9}$
7) $1\frac{5}{6}$ 8) $\frac{17}{24}$ 9) $\frac{1}{2}$ 10) 3

Go On Answers will vary

Partial Possibilities 6

1) $\frac{1}{4}, \frac{1}{3}, \frac{3}{8}, \frac{5}{6}$ 2) $1\frac{1}{6}$ 3) $\frac{1}{4}$ 4) $\frac{17}{24}$ 5) $2\frac{2}{3}$ 6) $1\frac{3}{10}$
7) $1\frac{3}{4}$ 8) $\frac{2}{15}$ 9) $\frac{5}{9}$ 10) $2\frac{2}{5}$

Go On One possible answer $\frac{14}{15}$ (numerator is one less than denominator)

p. 88 ***If...Then...***

1) 1 2) $2\frac{1}{2}$ 3) $4\frac{1}{2}$ 4) 2 5) 1 6) 2 7) 3 8) $\frac{1}{2}$
9) 6 10) $1\frac{1}{2}$ 11) $7\frac{1}{2}$ 12) $4\frac{1}{2}$ 13) $\frac{1}{8}$ 14) 1 15) $\frac{5}{8}$ 16) $1\frac{1}{2}$
17) 4 18) 12 19) 6 20) 2 21) $\frac{1}{6}$ 22) $\frac{1}{3}$ 23) $1\frac{1}{3}$ 24) $\frac{5}{6}$
25) $B = \frac{1}{6}$; $C = \frac{2}{3}$ or $\frac{4}{6}$; $D = \frac{5}{6}$ 26) $B = \frac{1}{4}$; $C = 1$; $D = 1\frac{1}{4}$ 27) $B = \frac{2}{3}$; $C = 2\frac{2}{3}$; $D = 3\frac{1}{3}$

p. 89 ***Digits to Fractions I***

1) $\frac{1}{4} + \frac{2}{5}$ 2) $\frac{5}{4} - \frac{1}{2}$ 3) $\frac{1}{2} + \frac{4}{5}$ 4) $\frac{1}{5} + \frac{2}{4}$ 5) $\frac{4}{5} - \frac{1}{2}$ or $\frac{2}{4} - \frac{1}{5}$ 6) $\frac{5}{2} - \frac{1}{4}$
7) $\frac{2}{8} + \frac{3}{5}$ 8) $\frac{8}{5} - \frac{2}{3}$ 9) $\frac{8}{5} + \frac{2}{3}$ 10) $\frac{5}{8} + \frac{2}{3}$ 11) $\frac{3}{5} - \frac{2}{8}$ 12) $\frac{8}{2} - \frac{5}{3}$

p. 90 ***Digits to Fractions II***

Possible answers.

1) $\frac{1}{4} \times \frac{2}{3}$ or $\frac{1}{3} \times \frac{2}{4}$ 2) $\frac{1}{2} \times \frac{3}{4}$ or $\frac{1}{4} \times \frac{3}{2}$ 3) $\frac{3}{2} \div \frac{1}{4}$ or $\frac{4}{2} \div \frac{1}{3}$ 4) $\frac{1}{3} \times \frac{4}{2}$ or $\frac{1}{2} \times \frac{4}{3}$
5) $\frac{2}{4} \div \frac{1}{3}$ or $\frac{3}{4} \div \frac{1}{2}$ 6) $\frac{2}{3} \div \frac{1}{4}$ or $\frac{4}{3} \div \frac{1}{2}$ 7) $\frac{1}{4} \times \frac{6}{5}$ or $\frac{6}{4} \times \frac{1}{5}$ 8) $\frac{1}{4} \times \frac{5}{6}$ or $\frac{1}{6} \times \frac{5}{4}$
9) $\frac{6}{1} \div \frac{4}{5}$ or $\frac{5}{4} \div \frac{1}{6}$ 10) $\frac{1}{5} \times \frac{4}{6}$ or $\frac{4}{5} \times \frac{1}{6}$ 11) $\frac{4}{5} \div \frac{1}{6}$ or $\frac{6}{5} \div \frac{1}{4}$ 12) $\frac{4}{6} \div \frac{1}{5}$ or $\frac{5}{1} \div \frac{6}{4}$

p. 91 ***Fraction Choices I***

1) $\frac{3}{8} + \frac{1}{8} + \frac{1}{4}$ 2) $\frac{1}{2} + \frac{1}{4} - \frac{1}{8}$ 3) $\frac{1}{2} - \frac{3}{8} + \frac{1}{4}$ 4) $\frac{1}{2} - \frac{3}{8} - \frac{1}{8}$ or $\frac{3}{8} - \frac{1}{4} - \frac{1}{8}$ 5) $\frac{1}{2} + \frac{1}{4} + \frac{3}{8}$ 6) $\frac{1}{2} - \frac{3}{8} + \frac{1}{8}$
7) $\frac{1}{2} - \frac{1}{8} - \frac{1}{4}$ 8) $\frac{3}{8} + \frac{1}{8} - \frac{1}{2}$ or $\frac{1}{4} + \frac{1}{8} - \frac{3}{8}$ 9) $\frac{2}{3} + \frac{1}{3} + \frac{1}{2}$ 10) $\frac{2}{3} + \frac{1}{2} - \frac{1}{6}$ 11) $\frac{2}{3} - \frac{1}{2} - \frac{1}{6}$ 12) $\frac{2}{3} - \frac{1}{2} + \frac{1}{6}$ or $\frac{1}{2} - \frac{1}{3} + \frac{1}{6}$
13) $\frac{2}{3} + \frac{1}{2} + \frac{1}{6}$ 14) $\frac{2}{3} - \frac{1}{3} - \frac{1}{6}$ 15) $\frac{1}{3} + \frac{2}{3} - \frac{1}{6}$ or $\frac{2}{3} + \frac{1}{2} - \frac{1}{3}$ 16) $\frac{1}{2} - \frac{1}{6} + \frac{1}{3}$

p. 92 ***Fraction Choices II***

1) $\frac{1}{4} \times \frac{1}{2} + \frac{1}{8}$ 2) $\frac{1}{4} \times \frac{3}{4} - \frac{1}{8}$ 3) $\frac{1}{4} \times \frac{3}{4} \times \frac{1}{2}$ 4) $\frac{1}{4} \times \frac{1}{2} - \frac{1}{8}$ 5) $\frac{1}{8} + (\frac{3}{4} \times \frac{1}{2})$ 6) $\frac{3}{4} \times \frac{1}{2} - \frac{1}{4}$
7) $\frac{1}{2} \times \frac{1}{4} \times \frac{1}{8}$ 8) $\frac{3}{4} + (\frac{1}{4} \times \frac{1}{2})$ 9) $\frac{1}{2} \times \frac{1}{3} + \frac{5}{6}$ 10) $\frac{2}{3} \times \frac{5}{6} - \frac{1}{3}$ 11) $\frac{1}{2} \times \frac{2}{3} \times \frac{5}{6}$ 12) $(\frac{2}{3} + \frac{1}{2}) \times \frac{1}{3}$
13) $\frac{1}{3} + (\frac{1}{2} \times \frac{2}{3})$ 14) $\frac{1}{2} \times \frac{2}{3} - \frac{1}{3}$ 15) $\frac{1}{2} \times \frac{2}{3} + \frac{5}{6}$ 16) $(\frac{5}{6} - \frac{1}{2}) \times \frac{2}{3}$

p. 101 ***Partial Possibilities 7***

1) 16.4 2) 0.375 3) $\frac{1}{2}, \frac{2}{3}$, .7 4) 6.23 + 58.7 + 5.07 5) 41.45
6) 39.84 7) 0.17 8) 9.1 9) 3.78 10) 1.8

Go On 14.22, 16.82, 19.42; numbers increase by 2.6

Partial Possibilities 8

1) 5.36 2) 0.8 3) .18, $\frac{1}{5}$, $\frac{1}{3}$ 4) 41.9 + .44 + 27.66 5) 10.15
6) 1.72 7) 0.26 8) 4 9) 1.6 10) 230
Go On 98.5 + 6.4 = 104.9

p. 102 ***Partial Possibilities 9***

1) 11.8 2) 0.875 3) $\frac{2}{5}$, $\frac{1}{2}$, .51 4) 7.16 + 42.5 + 15.34 5) 71.07
6) 22.57 7) 0.44 8) 7.2 9) 0.18 10) 0.5
Go On Possible answers: 7.8 × 4.6; 78 × .46; 46 × .78

Partial Possibilities 10

1) 7.21 2) .22 3) $\frac{1}{3}$, .37, .4 4) 32.08 + 8.62 + 79. 3 5) 33.4
6) 2.76 7) 0.28 8) 2.4 9) 0.135 10) 0.4
Go On 18.7, 22.3, 25.9 numbers increase by 3.6

p. 103 ***Partial Possibilities 11***

1) 9.07 2) 0.625 3) $\frac{3}{8}$, .5, $\frac{2}{3}$ 4) 120.4 + 3.97 + 25.63 5) 53.88
6) 0.4 7) 0.30 8) 12 9) 0.105 10) 30
Go On 4.26 – 3.7 = 0.56

Partial Possibilities 12

1) 12.02 2) 0.83 3) $\frac{1}{2}$, .72, $\frac{5}{6}$ 4) 52.7 + 40.81 + 6.49 5) 5.33
6) 4.53 7) 0.35 8) 10.8 9) 1.012 10) 6
Go On 96.3 × 7.5 = 722.25

p. 110 ***Identifying Missing Decimals*** Order of addends and/or factors may vary.

1) 8.7 + 74.5 2) 68.8 3) 68.8 + 23.4 4) 56.2 5) 56.2; 23.4 6) 56.2; 23.4
7) 47.1 8) 74.5; 47.1 9) 68.8 – 47.1 10) 56.2 + 47.1 11) 23.4 12) 68.8
13) 56.2 14) 59.24 + 34.7 15) 110.83 – 92.6 16) 46.1 + 6.9 17) 70.5 – 59.24 18) 83.38; 59.24
19) 83.38 + 46.1 20) 92.6 – 46.1 21) 70.5; 34.7 22) 92.6 + 46.1 23) 110.83 – 83.38 24) 70.5 × 6.9
25) 92.6 × 34.7 26) 70.5; 46.1 27) 34.7 × 70.5 28) 6.9; 46.1 29) 34.7; 46.1 30) 92.6 × 6.9
31) 6.9; 34.7 32) 46.1; 92.6

p. 111 ***Fitting Decimals I*** Answers may vary. Samples given. For problems 13–24 terms may vary although the final answer should be as shown.

1) 246.3 2) 62.43 3) $3.2 < 4.3 < 6.8$ 4) $4.6 > 2.4 < 3.5$
5) $1.3 > 1.28 > 0.64$ 6) $0.2 < 0.54 < 0.63$ 7) 3.4 + 0.75 8) 5.75 – 0.34
9) $8.3 + 0.45 < 9.7$ 10) $7.5 - 0.4 > 6.3$ 11) $75.4 + 7.5 > 14.3$ 12) $0.7 - 0.34 > 0.25$
13) 97.7 + 6.3 = 104 14) 49.6 + 7.35 = 56.95 15) 9.6 + 0.73 + 7.69 = 18.02
16) 97.6 – 36.9 = 60.7 17) 9.27 – 0.36 = 8.91 18) 9.46 – 3.33 + 0.75 = 6.88
19) 24.5 – 7.6 = 16.9 20) 34.7 + 26.8 = 61.5 21) 4.76 – 4.72 = 0.04 22) 24.3 – 5.76 = 18.54
23) 7.5 + 0.24 – 6.9 = 0.84 24) 7.7 – 6.64 + 0.28 = 1.34

p. 112 ***Fitting Decimals II*** Answers to problems 1–6 may vary. Samples are given.

1) 74.6 × 92 2) 27.9 × 24 3) 54.9 × 32.7 4) 9.82 × 7.4
5) 972.4 6) 4.3 × 27.9 7) 27.9 × 47.6 8) 942.4 × 7.5
9) 7.5 × 8.4 = 63 10) 84.7 × 7.5 = 635.25 11) 75.4 × 0.8 = 60.32 12) 0.75 × 8.4 = 6.3
13) 3.6 × 5.8 = 20.88 14) 56.8 × 0.3 = 17.04 15) 35.8 × 6.6 = 236.28 16) 4.68 × 3.5 = 16.38
17) 32.8 18) 76.8 ÷ 3.2 19) 110.4 ÷ 0.6 = 23 × 8 20) 100.8 ÷ 28 = 3.6
21) 23.6 ÷ 8 22) 215.8 ÷ 26 = 8.3

p. 119 ***Partial Possibilities 13***

1) 20% 2) $\frac{4}{5}$ 3) 3 4) 7.7 5) 5% 6) 40%
7) $\frac{1}{3}$, 37%, 0.40 8) $27 9) $\frac{3}{7} < 65\%$ or $\frac{3}{6} < 75\%$ 10) $76\% > \frac{3}{4}$
Go On 40% of 50 is greater; proofs will vary

Partial Possibilities 14

1) 75% 2) $\frac{7}{25}$ 3) 20 4) 4.5 5) 35% 6) 28% 7) 0.59, 72%, $\frac{7}{8}$

8) $24 9) $\frac{5}{2} > 49\%, \frac{5}{9} > 42\%, \frac{9}{2} > 45\%$, or $\frac{9}{5} > 42\%$ 10) $25\% < \frac{3}{9}, 29\% < \frac{3}{5}, 59\% < \frac{3}{2}$, or $95\% < \frac{3}{2}$

Go On Answers will vary.

p. 120 *Partial Possibilities 15*

1) 70% 2) $\frac{1}{5}$ 3) 12 4) 9.6 5) 30% 6) 15%

7) 32%, $\frac{5}{9}$, 0.62 8) $56 9) $\frac{3}{8} < 52\%$ or $\frac{3}{5} < 82\%$ 10) $85\% > \frac{3}{5}$

Go On 30% of 120 is greater; proofs will vary.

Partial Possibilities 16

1) 25% 2) $\frac{11}{20}$ 3) 24 4) 7.8 5) 55% 6) 17%

7) 0.35, $\frac{3}{7}$, 53% 8) $8.50 9) $\frac{2}{6} < 85\%$ or $\frac{2}{8} < 65\%$ 10) $68\% > \frac{2}{3}$ or $86\% > \frac{2}{3}$

Go On Answers will vary.

p. 121 *Partial Possibilities 17*

1) 50% 2) $\frac{2}{5}$ 3) 45 4) 13.3 5) 15% 6) 5%

7) $\frac{1}{6}$, 24%, .33 8) $30 9) $\frac{3}{7} < 48\%$ or $\frac{3}{4} < 78\%$ 10) $34\% < \frac{7}{5}$, 43 % $< \frac{7}{5}$, or $47\% < \frac{3}{5}$

Go On 40% of 90 is greater; proofs will vary.

Partial Possibilities 18

1) 40% 2) $\frac{3}{4}$ 3) 15 4) 10.8 5) 55% 6) 23%

7) 0.58, 71%, $\frac{8}{9}$ 8) $44 9) $\frac{5}{8} < 64\%$ or $\frac{4}{8} < 65\%$ 10) $48\% < \frac{3}{5}$ or $58\% < \frac{3}{4}$

Go On Answers will vary.

p. 129 *Finding Tic–Tac–Toes*

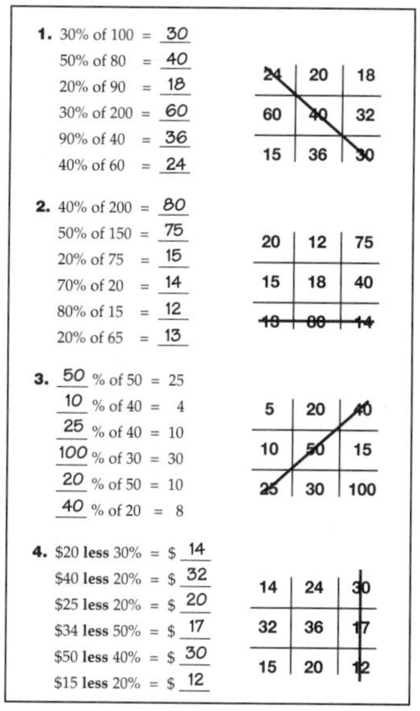
1. 30% of 100 = 30
50% of 80 = 40
20% of 90 = 18
30% of 200 = 60
90% of 40 = 36
40% of 60 = 24

24	20	18
60	40	32
15	36	30

2. 40% of 200 = 80
50% of 150 = 75
20% of 75 = 15
70% of 20 = 14
80% of 15 = 12
20% of 65 = 13

20	12	75
15	18	40
13	80	14

3. 50 % of 50 = 25
10 % of 40 = 4
25 % of 40 = 10
100 % of 30 = 30
20 % of 50 = 10
40 % of 20 = 8

5	20	40
10	50	15
25	30	100

4. $20 less 30% = $ 14
$40 less 20% = $ 32
$25 less 20% = $ 20
$34 less 50% = $ 17
$50 less 40% = $ 30
$15 less 20% = $ 12

14	24	30
32	36	17
15	20	12

Trivia: 17 trees

p. 130 *Figuring Percents Practice I*

Answers will vary.

1) 25% of 60 = 15 or 75% of 60 = 45 2) 75% of 80 = 60 3) 75% of 60 = 45
4) 25% of 40 = 10, 25% of 60 = 15, 25% of 80 = 20, or 50% of 40 = 20 5) 60% of 30 = 18
6) 60% of 70 = 42 7) 10% of 30 = 3, 10% of 50 = 5, 10% of 70 = 7, or 30% of 30 = 9
8) 60% of 50 = 30 9) 60% of 80 = 48 10) 40% of 75 = 30 or 40% of 80 = 32
11) 15% of 20 = 3 or 40% of 20 = 8 12) 60% of 75 = 45

p. 131 ***Figuring Percents Practice II***

Answers will vary.

1) 85% of 400 = 340 2) 35% of 500 = 175 3) 55% of 400 = 220, 35% of 700 = 245, or 55% of 500 = 275
4) 85% of 700 = 595 5) 90% of 750 = 675 6) 65% of 1000 = 650 or 90% of 750 = 675
7) 40% of 1000 = 400, 40% of 750 = 300, or 90% of 1000 = 900
8) 90% of 1000 = 900 9) 70% of 720 = 504 or 60% of 960 = 576
10) 30% of 360 = 108, 30% of 480 = 144, 30% of 720 = 216, or 60% of 360 = 216
11) 30% of 960 = 288, 60% of 480 = 288, 70% of 360 = 252, or 70% of 480 = 336 12) 60% of 720 = 432

p. 132 ***Fitting Percents***

Answers will vary. Samples given.

1) 0.65 > 38% 2) 35% < 0.86 3) $\frac{1}{3}$ < 56% 4) $\frac{5}{6}$ > 38%
5) 0.97 + 85% or 0.87 + 95% = 182% 6) 98% – 0.35 = 63% 7) $\frac{9}{3}$ – 57% = 243%
8) $\frac{9}{3}$ + 0.87 = 387% 9) 0.36 + 48% = 84% 10) 46% – 0.39 = 7%
11) 84% – $\frac{3}{4}$ = 9% 12) $3\frac{4}{8}$ + 0.68 = 418% 13) 69% < 0.75 > 24% 14) 0.67 > $\frac{2}{4}$ > 19%
15) $\frac{2}{4}$ > 39% < 0.67 16) 0.27 < $\frac{4}{6}$ < 89% 17) 0.85 + 63% + $\frac{2}{4}$ = 198% 18) 0.69 – 53% + $\frac{8}{4}$ = 216%

p. 139 ***Number Explorations 1***

1) Possible factors: 1, 2, 4, 5, 10, 20 2) 6 3) 45 4) 24 5) Possible prime numbers: 2, 3, 5, 7
6) $x = 6$ 7) $x = 8$ 8) $x = 24$ 9) 17 10) 11

Go on

x	y	$x + y$	$x - y$
30	18	48	12
50	20	70	30

Number Explorations 2

1) Possible factors: 1, 2, 4, 8, 16 2) 2 3) 42 or 49 4) 36 5) 31, 37
6) $x = 6$ 7) $x = 7$ 8) $x = 5$ 9) 34 10) 16

Go on 9, 25, 49. Riddles will vary.

p. 140 ***Number Explorations 3***

1) Possible factors: 1, 2, 3, 6, 9, 18 2) 4 3) 36 4) 48 5) 11, 13
6) $x = 5$ 7) $x = 8$ 8) $x = 18$ 9) 27 10) 20

Go on

a	b	$a + b$	$a - b$
60	12	72	48
40	25	65	15

Number Explorations 4

1) Possible factors: 1, 2, 3, 4, 6, 8, 12, 24 2) 2 3) 48 or 56 4) 45 5) 17, 19
6) $x = 10$ 7) $x = 8$ 8) $x = 4$ 9) 34 10) 1

Go on 30, 42, 48. Riddles will vary.

p. 141 ***Number Explorations 5***

1) Possible factors: 1, 2, 3, 4, 6, 12 2) 3 3) 24 or 28 4) 24 5) 23, 29
6) $x = 3$ 7) $x = 5$ 8) $x = 16$ 9) 48 10) 41

Go on

c	d	$c + d$	$c - d$
80	15	95	65
60	30	90	30

Number Explorations 6

1) Possible factors: 1, 2, 3, 4, 6, 9, 12, 18, 36 2) 8 3) 36 or 45 4) 72 5) 41, 43, 47
6) $x = 6$ 7) $x = 7$ 8) $x = 6$ 9) 37 10) 43

Go on 12, 24, 48. Riddles will vary.

p. 147 ***Number Riddle Arrangements***

1)

1	10
7	3

2)

5	7
10	2

3)

1	5	8
6	4	2

4)

11	2	12
6	9	3

p. 150 ***Discover Trivia by Identifying Unknowns***

S) $x = 11$ U) $n = 4$ L) $x = 9$ O) $n = 13$ V) $x = 15$ I) $n = 7$ N) $x = 8$
I) $n = 12$ N) $x = 6$ A) $n = 2$ I) $x = 5$ E) $n = 16$ E) $x = 10$ L) $n = 14$

The most landed-on square is Illinois Avenue.